LATIN AMERICAN JOURNALISM

COMMUNICATION
TEXTBOOK SERIES
Jennings Bryant—Editor

Journalism
Maxwell McCombs—Advisor

ELECTROANALYTICAL CHEMISTRY

VOLUME 17

ELECTROANALYTICAL CHEMISTRY

A SERIES OF ADVANCES

Edited by
ALLEN J. BARD

DEPARTMENT OF CHEMISTRY
UNIVERSITY OF TEXAS
AUSTIN, TEXAS

VOLUME 17

CRC Press
Taylor & Francis Group
Boca Raton London New York

CRC Press is an imprint of the
Taylor & Francis Group, an **informa** business

First published 1991 by Marcel Dekker, Inc.

Published 2009 by CRC Press
Taylor & Francis Group
6000 Broken Sound Parkway NW, Suite 300
Boca Raton, FL 33487-2742

© 1991 by Taylor & Francis Group, LLC
CRC Press is an imprint of Taylor & Francis Group, an Informa business

First issued in paperback 2019

No claim to original U.S. Government works

ISBN 13: 978-0-367-45074-8 (pbk)
ISBN 13: 978-0-8247-8409-6 (hbk)

Visit the Taylor & Francis Web site at
http://www.taylorandfrancis.com

and the CRC Press Web site at
http://www.crcpress.com

The Library of Congress Catalogued the First
Issue of This Title as Follows:

Electroanalytical chemistry: a series of advances, v. 1

   New York, M. Dekker, 1966-
         v. 23 cm.
   Editor: 1966- A. J. Bard

   1. Electrochemical analysis-Addresses, essays, lectures
   1. Bard, Allen J., ed.

QD115E499   545.3   66-11287

Library of Congress
ISBN 0-8247-8409-X (v.17)

Introduction to the Series

This series is designed to provide authoritative reviews in the field of modern electroanalytical chemistry defined in its broadest sense. Coverage will be comprehensive and critical. Enough space will be devoted to each chapter of each volume so that derivations of fundamental equations, detailed descriptions of apparatus and techniques, and complete discussions of important articles can be provided, so that the chapters may be useful without repeated reference to the periodical literature. Chapters will vary in length and subject area. Some will be reviews of recent developments and applications of well-established techniques, whereas others will contain discussion of the background and problems in areas still being investigated extensively and in which many statements may still be tentative. Finally, chapters on techniques generally outside the scope of electroanalytical chemistry, but which can be applied fruitfully to electrochemical problems, will be included.

Electroanalytical chemists and others are concerned not only with the application of new and classical techniques to analytical problems, but also with the fundamental theoretical principles upon which these techniques are based. Electroanalytical techniques are proving useful in such diverse fields as electro-organic synthesis, fuel cell studies, and radical ion formation, as well as with such problems as the kinetics and mechanisms of electrode reactions, and the effects of electrode surface phenomena, adsorption, and the electrical double layer on electrode reactions.

It is hoped that the series will prove useful to the specialist and nonspecialist alike—that it will provide a background and a

starting point for graduate students undertaking research in the
areas mentioned, and that it will also prove valuable to practic-
ing analytical chemists interested in learning about and applying
electroanalytical techniques. Furthermore, electrochemists and
industrial chemists with problems of electrosynthesis, electro-
plating, corrosion, and fuel cells, as well as other chemists wish-
ing to apply electrochemical techniques to chemical problems, may
find useful material in these volumes.

A. J. B.

Contributors to Volume 17

BARBARA BITTINS-CATTANEO   Institute of Physical Chemistry, University of Bonn, Bonn, Federal Republic of Germany

DANIEL A. BUTTRY   University of Wyoming, Laramie, Wyoming

EDUARDO CATTANEO   Institute of Physical Chemistry, University of Bonn, Bonn, Federal Republic of Germany

PETER KÖNIGSHOVEN   Institute of Physical Chemistry, University of Bonn, Bonn, Federal Republic of Germany

RICHARD L. McCREERY   Department of Chemistry, The Ohio State University, Columbus, Ohio

GERALDINE L. RICHMOND   Department of Chemistry, University of Oregon, Eugene, Oregon

WOLF VIELSTICH   Institute of Physical Chemistry, University of Bonn, Bonn, Federal Republic of Germany

Contents of Volume 17

Contents of Other Volumes

VOLUME 1

AC Polarograph and Related Techniques: Theory and Practice,
    Donald E. Smith
Applications of Chronopotentiometry to Problems in Analytical
    Chemistry, Donald G. Davis
Photoelectrochemistry and Electroluminescence, Theodore Kuwana
The Electrical Double Layer, Part I: Elements of Double-Layer
    Theory, David M. Mohilner

VOLUME 2

Electrochemistry of Aromatic Hydrocarbons and Related Substances,
    Michael E. Peover
Stripping Voltammetry, Embrecht Barendrecht
The Anodic Film on Platinum Electrodes, S. Gilaman
Oscillographic Polarography at Controlled Alternating Current,
    Michael Heyrovsky and Karel Micka

VOLUME 3

Application of Controlled-Current Coulometry to Reaction Kinetics,
    Jiri Janata and Harry B. Mark, Jr.
Nonaqueous Solvents for Electrochemical Use, Charles K. Mann

VOLUME 8

VOLUME 9

VOLUME 10

VOLUME 11

VOLUME 12

APPLICATIONS OF THE QUARTZ
CRYSTAL MICROBALANCE TO ELECTROCHEMISTRY

Daniel A. Buttry

University of Wyoming
Laramie, Wyoming

I. INTRODUCTION

The quartz crystal microbalance (QCM) is a piezoelectric device capable of extremely sensitive mass measurements. It oscillates in a mechanically resonant shear mode by application of an alternating, high frequency electric field using electrodes which are usually deposited on both sides of the disk. Sauerbrey was the first to recognize that these devices could be used to measure mass changes at the crystal surface [1]. The mass sensitivity arises from a dependence of the oscillation frequency on the total mass of the (usually disk-shaped) crystal, its electrodes, and any materials present at the electrode surface. These devices have been used for many years to measure the masses of thin films in various types of deposition processes [2]. To the extent that the density of the deposit is known, the thickness may be calculated, so these devices are used in a number of commercial film thickness monitors. Until about 10 years ago it was thought that these crystals would not oscillate in liquids due to excessive energy loss to the solution from viscous effects. At that time Konash and Bastiaans [3] and Nomura [4] demonstrated the use of the QCM in the liquid environment for the determination of mass changes at the crystal surface. These reports made clear the potential utility of the QCM for accurate, in situ determinations of extremely small mass changes of the crystal electrodes or films deposited on them.

The use of the QCM in an electrochemical context to monitor mass changes at electrodes was first demonstrated by Jones and Meiure [5,6], who showed that trace metal determinations were possible by plating the metals onto a QCM electrode and measuring the change in resonant frequency of the crystal in air following its removal from the electrochemical cell. The first in situ application of the QCM to electrochemical problems was by Nomura and co-workers [7,8], who used it to determine Cu(II) and Ag(I) by electrodeposition. (Such in situ applications of the QCM to electrochemical systems will be distinguished from nonelectrochemical applications by referring to the former as EQCM (electrochemical QCM) experiments.) When used as an in situ technique for measuring mass changes at electrode surfaces, one of the EQCM electrodes is used simultaneously to provide for the alternating electric field which drives the oscillation of the crystal and as the working electrode in the electrochemical cell. Thus, the EQCM experiment involves the measurement of the various electrochemical parameters, such as potential, current, and charge, at one of the EQCM electrodes and the simultaneous measurement of the oscillation frequency of the piezoelectric crystal from which, in favorable cases, minute mass changes at the electrode may be inferred. Various instrumental approaches to this have been used, which will be described in a later section.

Since the pioneering publications of Nomura appeared, the EQCM has been used to study monolayer and multilayer depositions and dissolutions, mass transport in polymer films on electrodes, corrosion processes at electrodes, electroless depositions, and mass changes caused by protein adsorption at electrodes. In addition, many other applications are on the horizon. It is the purpose of this chapter to describe the use of the EQCM to study problems of interest to the electrochemical community, to give information about the experimental aspects of its in situ use, and to discuss its potential for future application. The treatment of the many EQCM studies which have appeared to date is not exhaustive, but rather was chosen to serve as a basis for discussion of the important issues in the use of the EQCM. A detailed review of the piezoelectric effect in quartz crystals is not presented here, as this topic has been adequately discussed elsewhere (see Ref. 2 and references therein). Also, the use of the QCM for other (nonelectrochemical) analytical

applications related to the determination of mass changes at surfaces in liquids or gases, while of considerable current interest,
is not discussed in detail here, except to the extent that such
results impact on electrochemical studies.

II.   EXPERIMENTAL METHODS
A.   Instrumentation and Materials
1.   Crystals

Nearly all reported applications of the QCM have employed alpha
quartz crystals because of the superior mechanical and piezoelectric properties of alpha quartz. The QCM transducers are prepared (usually by the manufacturer) by cutting the desired parts
from large single crystals of alpha quartz at certain angles with
respect to the crystalline axes of the quartz crystal [9]. For
QCM applications, the AT, BT, and SC (stress-compensated)
cuts [9] have been most frequently employed. AT-cut crystals
are particularly popular because they can be cut to give nearly
zero temperature coefficients (the proportionality constant relating the oscillation frequency of a crystal in vacuum to its temperature) at one or two temperatures [9]. However, as will be
discussed below, temperature effects for crystals immersed in
liquids far outweigh the intrinsic temperature dependence of the
crystals themselves and must be taken into account in most electrochemical applications. SC-cut crystals would seem to offer
certain advantages in terms of the elimination of stress-induced
frequency changes (vide infra) but have not as yet been investigated in EQCM applications. In the remainder of this chapter,
the discussion will be restricted to AT-cut crystals unless otherwise specified.

QCM crystals are typically employed as thin, disk-shaped
transducers which oscillate in a pure shear mode when an alternating electric field of the proper frequency is applied across
the disk, i.e., the electric field lines are normal to the disk

surface. Figure 1 shows an edge view of a QCM disk with an exaggerated view of the shear distortion from the oscillation. The designation of the oscillation as a pure shear mode indicates that the motion of the disk surface is precisely parallel to the disk face. Typical frequencies for these oscillations in EQCMs are from 1 to 10 megahertz (MHz).

Aside from the thickness shear mode described above, other modes of vibration can also be excited by the applied electric field. While most other modes are only weakly coupled to the shear mode, the flexural modes are sometimes strongly coupled [10]. When the frequencies of these flexural modes are near that of the shear mode, deviations in resonant frequency can occur such that the response of the QCM to mass changes is no longer linear. Fortunately, it is usually possible to eliminate this source of error by proper design of the crystal and its electrodes. This will be described further below.

For shear mode oscillation, there are several frequencies which correspond to resonant conditions. The distinctions between these and their relevance will be described later. For the present purposes, the resonant frequency of the crystal may be identified as the frequency of maximum displacement of the crystal surface (for a constant driving voltage). This condition corresponds to the establishment of a standing acoustic wave within the bulk of the crystal with a node existing in the center

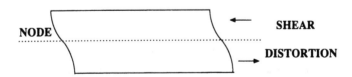

FIG. 1. Edge view of QCM disk showing shear deformation. The disk thickness and shear deformation have been exaggerated for clarity.

of the disk (see Fig. 1) and the antinodes at the two surfaces. The resonant frequency can thus be seen to be related to the thickness of the crystal through the following equation:

$$t_q = \frac{v_q}{2f_o} = \frac{\lambda_q}{2} \tag{1}$$

where v_q is the velocity of the acoustic wave in quartz, f_o is the resonant frequency, t_q is the thickness, and λ_q is the wavelength of the acoustic wave in quartz. This equation shows the reciprocal dependence of f_o on the crystal thickness. For an AT-cut crystal, v_q = 3340 m/s [11], which gives t_q = 334 μm and λ_q = 668 μm for a crystal with a resonant frequency of 5 MHz. Higher frequency, odd harmonics may also be excited. These will be considered further below in regards to the higher mass sensitivity obtained at these higher frequencies.

Figure 2 shows a top view of a QCM disk with electrodes in what is sometimes called a keyhole pattern. In this pattern the central area of the crystal is sandwiched between two concentric electrodes on opposite sides of the crystal disk, with the contact "flags" extending towards opposite edges of the crystal. The crystals used in our laboratories are one inch in diameter, with a concentric disk electrode of area 0.28 cm^2. Smaller diameter crystals may also be used [12]. Small crystals have the advantage of being less expensive, but they suffer from the serious drawback that the structures used to mount the crystal are closer to the piezoelectrically active area, so the influence of stress from the mounting can become a problem [13]. Larger crystals allow for the mounting structures to be farther from the region of the crystal undergoing displacement (which is nearly confined to that part of the crystal sandwiched between the two circular pads in the center of the disk [1,14]).

As stated above, the dimensions of the electrodes and the crystal disk have a strong influence on the coupling of other

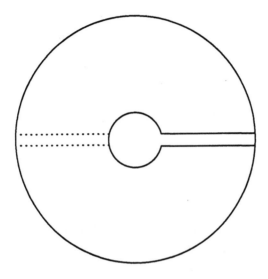

FIG. 2.   Top view of QCM disk with vapor deposited electrodes.
Disk diameter—1 inch.   Area of central circular pad—0.28 cm^2.
Width of rectangular flag for external connection—2 mm.   The
thin film metal electrodes are ca. 300 nm thick.

modes to the thickness shear mode.   This coupling represents a
source of spurious frequency changes which must be addressed
for optimum performance of the EQCM.   A recent discussion of
the frequency, amplitude, and shape of vibrations in quartz
crystals described many of the design criteria which provide
for the suppression of unwanted modes [15].   The critical dimensions are the thickness of the quartz disk (t_q), the diameter of the crystal (d_c), and the diameter of the concentric
electrode pad (d_e).   For a ratio of d_c/t_q of greater than 50,
unwanted modes may be suppressed by 40 dB when the ratio of
d_e/t_q is larger than 18 [15].   The values of these ratios for
the design in Fig. 2 are 85 and 20, respectively, well within
the acceptable ranges for excitation of the thickness shear mode
without significant coupling to other modes.   Following these

design criteria, the minimum value of d_c for a 5 MHz crystal (t_q = 3 × 10^{-2} cm) is 1.5 cm, and the minimum value of d_e is 0.54 cm. These calculations do not take into consideration the "flag" which extends from the edge of the circular part of the electrode to the edge of the crystal disk, but it has generally been found that the flag has minimal influence on oscillator stability.

Another way of considering the problems of coupling to unwanted modes or to the structures used to mount the crystal is with the concept of energy trapping [9,15]. Energy trapping refers to the fact that most of the energy contained in the mechanical oscillation of the crystal can be confined to the electroded region of the crystal if this region has a resonant frequency which is significantly lower than that of the nonelectroded region. This can be easily effected by using rather thick electrodes (with thicknesses greater than, say, 100 nm) or by contouring one or both faces of the crystal disk so that the thickness of the electroded region is significantly different from that of the rest of the disk [9].

The two predominant types of AT-cut crystals used for mass measurements are the so-called plane and the plano-convex. Plane crystals have both disk surfaces parallel to within ca. 1 μm, depending on the manufacturer. Plano-convex crystals have one side of the disk flat and the other ground to have a very slight radius of curvature. A typical range of radius of curvature for a 1-inch diameter, 5 MHz crystal would be 10–50 cm. For plano-convex crystals the region of the crystal which undergoes displacement during the oscillation (and therefore senses mass changes) is confined completely to the area defined by the electrodes (or by the smaller of the two electrodes if they have different sizes) [13]. For plane crystals this region extends somewhat past the edge of the electrode, onto the face of the quartz disk itself [1,13,15]. The degree of extension of this

region depends on the total mass loading of the crystal, with
the displacement being more completely confined to that part of
the crystal sandwiched between the electrodes for higher loading.
The amplitude of the displacement is known to depend on the
driving voltage [15] and probably also depends on the total mass
loading. For AT-cut crystals operating in the frequency range
of 1–10 MHz in the fundamental mode, the range of amplitudes
reported is ca. 10–100 nm, as determined by a wide variety of
optical methods as well as electron microscopy [1,15,16]. These
are fairly large displacements with respect to molecular dimensions, but it should be kept in mind that the relative velocities
and displacements of the electrode surface with respect to the
solution are much smaller. This is because the boundary layer
of solution very near to the surface is dragged along with the
surface at a similar velocity.

The distribution of the vibrational amplitude as a function
of radial distance from the center of the electrode is also known
from the same studies [15–17]. The amplitude is maximum in
the center of the disk and decays in an apparently Gaussian
fashion with distance, nearly reaching zero at the edge of the
electrode. This Gaussian distribution becomes considerably
sharper for plano-convex crystals as the radius of curvature is
decreased or with increasing harmonic number. For example,
for a 4 MHz crystal operating in the fundamental mode, the radial distance at which the amplitude is 1/e of its maximum value
decreases by 30% when the radius of curvature is decreased
from 50 to 6 cm [16]. Also, for a crystal having a radius of
curvature of 10 cm, this distance is decreased by 50% when the
crystal is operated in its third harmonic mode (12 MHz) as opposed to the fundamental mode (4 MHz) [16]. Thus, for plano-
convex crystals or for plane crystals operated in the third harmonic, the vibrational amplitude can be considered to be essentially zero at the edge of the electrode. For plane crystals

operated in the fundamental mode in vacuum or in a low density
gaseous environment, the amplitude distribution is known to be
more confined to the electrode area as mass loading is increased
[13].

Since the mass measurement made with quartz resonators is
actually an inertial measurement, the mass sensitivity is intimately
related to the vibrational amplitude distribution. There is ample
experimental evidence supporting a Gaussian mass sensitivity for
AT-cut crystals [1,13,15–19] with the maximum sensitivity at the
center of the electrode and zero sensitivity at the edges for
plano-convex or harmonically driven plane crystals. In one
particularly revealing study, the mass sensitivity distribution
for plane crystals was shown to become slightly more confined
to the electrode region as the mass loading was increased [13].
The differential mass sensitivity, c_f = df/dm, was shown to be
a function of radial distance from the center, while the integrated
mass sensitivity, C_f, given by Eq. (2) was shown to change
slightly at small loadings and then become relatively constant
for larger loadings [1,15–20]:

$$C_f = \int c_f \, dA \qquad\qquad\qquad (2)$$

where A is the area of the vibrational displacement.

The implication of this differential radial mass sensitivity for
EQCM applications is that the distribution of any mass changes
at the electrode surface must be known, whether or not it is
uniform. That the radial mass sensitivity changes with mass
loading might at first glance seem to cause problems (i.e., a
different proportionality constant relating frequency and mass
for different mass loadings). However, the evidence is that
most of the change occurs at fairly low mass loadings, with little
change in the radial mass sensitivity at higher loadings [13,19,
20]. Thus, since the mass loading from the vapor deposited
electrodes and from immersion of the crystal in a liquid is so

large, it would seem that this alone would serve the function of keeping the vibration amplitude confined to the electrode region, thereby bringing the value of C_f to its high-loading limit. This is an important consideration for precise, in situ mass measurements with these devices which has not yet been addressed experimentally.

A thin (ca. 2–5 nm) adhesion layer of either Cr or Si is usually deposited directly onto the quartz crystal to aid in the adhesion of the metal electrode. Spurious electrochemical responses can sometimes result if diffusion of the material from the adhesion layer to the electrode surface occurs. Au electrodes have been the most commonly used in EQCM studies because of the ease with which Au is evaporated. However, Cu, Pt, Ni, and other metals have also been employed. In principle, any type of material which can be deposited onto the surface of the underlying metal electrode, either by electrodeposition or from vacuum, can be used. The only limitations on the use of such materials are that they have good adhesion to the underlying electrode (which must be present to supply the alternating electric field) and the deposition must be carried out in such a way that the temperature of the quartz crystal does not exceed 573°C, above which alpha quartz loses its piezoelectric activity [21]. It is worth pointing out here that mass changes at the EQCM electrodes influence f_0 because these electrodes actually become a part of the composite resonator composed of the quartz crystal, its electrodes, any film deposited on the electrodes, and any liquid adjacent to the electrode (or deposit) surface which experiences shear forces. Thus, when the electrodes become delaminated due to poor adhesion of the underlayer, large, discontinuous changes in frequency occur which render a particular crystal useless.

Considering both convenience and cost, it is nearly essential that crystals be reused, i.e., cleaned and recoated with fresh

electrodes when a particular experiment is over. This can be
done if facilities for vacuum deposition are available. Either
thermal or sputtering techniques may be employed. To ensure
reproducible results careful attention to cleanliness must be paid,
both in the vacuum chamber and in the preparation of the crystal blanks prior to deposition. Normal methods for cleaning glass
parts may be used with the exception of basic baths, which will
cause etching.

Crystals may be obtained with either rough or smooth surfaces. Rough surfaces are the most common because they are
less expensive, but they suffer from a quantitatively unpredict-
able dependence of the absolute oscillation frequency in a liquid
on the trapping of the liquid in the pores on the surface [22,23].
The influence of surface roughness and deposit morphology on
the oscillation frequency will be discussed further below. Crystals having smooth surfaces are sometimes called overtone polished
crystals because the excitation of overtone (harmonic) modes requires a flatter surface due to the smaller wavelength of the
acoustic wave at higher frequencies. This type of crystal is
generally preferred because roughness effects then exist only
in the deposit (which might be of interest, depending on the
system studied) and do not arise from the underlying electrodes.

As stated above, EQCMs may be operated either in the fundamental mode or in any of the higher frequency, odd harmonics.
The advantage of higher frequency operation is a greater mass
sensitivity. However, since f_0 is inversely related to t_q, it becomes increasingly difficult to handle the extremely thin crystals
needed for high frequency operation in the fundamental mode.
Thus, a practical limit of about 15 MHz exists for the fundamental frequency of crystals which can be easily handled. Note
that a 15 MHz crystal is just slightly thicker than 100 μm. Use
of lower frequency crystals at odd harmonics is possible [24] at
the expense of slightly more complicated oscillator circuitry.

Another potential difficulty with the use of higher frequencies is that the viscous damping from the solution can more easily cause the oscillation to cease, depending on the circuitry used. A common cause of this would be roll-off in the gain of the oscillator circuit at the higher frequency. A good compromise used in our laboratories is the use of 5 MHz crystals, which offer acceptable (submonolayer) mass sensitivity, relative robustness, and stable oscillation in most viscous media (given the proper oscillation circuitry).

2. Cells

The method of mounting the crystals to the electrochemical cell is an important consideration because of possible influences of stress on the absolute frequency of oscillation, either in air or liquid, and because simple and rapid exchange of crystals is required when the thin film electrodes become damaged (which can occur from large potential excursions, aggresive chemical treatments, or mechanical abrasion). Several methods for mounting have been described [12,25–29], all of which make it possible to expose only one side of the EQCM disk (the working electrode side) to the electrolyte solution. This is usually necessary to prevent the two EQCM electrodes from being capacitively shunted by the solution, which can cause the cessation of oscillation [12]. It is possible to simply attach the crystals to an opening on the cell with some type of adhesive [12]. However, this method does not provide for easy replacement of the crystal when damage occurs. Another method which allows the use of glass cells is to sandwich the crystal between two o-rings in a vacuum o-ring joint which has been blown onto a standard H-cell [26]. In our laboratories a number 9 o-ring joint is used with the material of the o-ring chosen for compatibility with the solvent system under investigation. Finally, cells of Kel-F, Teflon, or other chemically inert materials can be employed with mounting again achieved by sandwiching between o-rings [25,27-29]. Large diameter crystals and o-rings seem to be best for keeping the mounting stresses away from the piezoelectrically active area. In all cases, the cells should be designed so that the oscillator circuitry is situated as close as possible to the crystal to minimize the length of the leads from the circuit to the crystal.

3. Oscillator Circuits

The oscillator circuit used determines in large measure the stability of the instrument and, therefore, the types of experiments that can be done. A key requirement of the circuit is that it provide sufficient gain to allow for oscillation of the crystal in a viscous medium. Three different designs have thus far been reported [12,24,25]. In one of the designs, the oscillation frequency of the EQCM crystal is measured with respect to a reference crystal which is external to the electrochemical cell [12]. Depending on the configuration used, the output can be sent to a frequency-to-voltage converter or a frequency counter. This design has the advantage of not requiring the use of a frequency counter. A disadvantage of this design would seem to be its failure to use a true ground for the EQCM working electrode. However, this system has been used to produce data on monolayer mass changes (perhaps the most challenging application for EQCM instrumentation) with very good signal-to-noise ratios. The second design [25] uses a modified Pierce-Miller oscillator with a high frequency 2N1711 transistor operating as an inverting amplifier. This is a more conventional design, similar to those used in many types of frequency control oscillator circuits [30]. The EQCM working electrode is at true ground, facilitating the connection of electrochemical circuitry, and some adjustment is provided to increase signal amplitude when viscous losses are severe.

Figure 3 shows the oscillator circuit used in the author's laboratory. This is essentially the same as that designed by Kanazawa [24], but it lacks the inductor-capacitor (LC) tuned

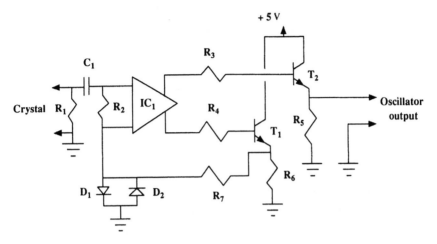

FIG. 3. Schematic of oscillator circuit. IC_1—MC1733; R_1—2.2 MΩ; R_2—200 Ω; R_3, R_4, R_5, R_6—180 Ω; R_7—220 Ω; C_1—0.01 μF; T_1, T_2—2N3904; D_1, D_2—HP 5082-2811 Schottky diodes.

element used to provide for oscillation at the third harmonic. This oscillator uses a Motorola MC1733 differential video amplifier as the central component, with a 2N3904 switching transistor in the feedback loop to provide for the signal inversion needed to achieve oscillation. The MC1733 is strapped for maximum gain (x400) and the transistor also provides significant gain (x50–200). The Schottky diodes act as limiters to prevent the crystal from damage due to large driving voltages. The values of the resistors in the feedback loop were chosen to provide for maximum stability of the oscillation frequency under conditions of viscous loading. The other output of the differential amplifier is sent to a frequency counter, which will be described below. A significant attribute of this circuit is that it provides a relatively constant ac voltage across the crystal (ca. 0.4 V peak to peak with a power supply voltage of 3.5 V) for a wide range of viscous loading conditions. This will be true so long as the gain is large enough for the diode limiters to be the elements determining the voltage in the feedback loop. Thus, the circuit is usually able to sustain crystal oscillation even under conditions of large viscous loading from thick, viscoelastic polymer films and highly viscous solutions. Most importantly, changes in this viscous loading do not appear to have strong influence on the characteristics of the oscillator circuit. In addition, the EQCM working electrode is at true ground, which conveniently allows the use of Wenking-style potentiostats. Several groups have reported the successful use of this circuit [24,26,29,31–33].

4. EQCM Apparatus

Figure 4 shows a schematic of the EQCM instrument used in the author's laboratory. The quartz crystal is mounted in the electrochemical cell with one electrode exposed. This electrode is maintained at true ground and is connected to the ground of the oscillator circuit and to the working electrode lead of the Wenking style potentiostat previously described [24]. The other crystal electrode is connected to the "hot" side of the oscillator circuit.

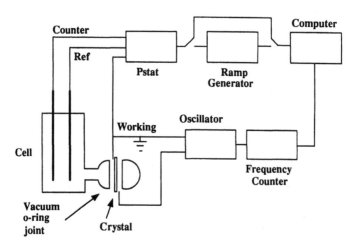

FIG. 4.   Schematic of EQCM apparatus.

These connections are made using small, flat-nosed alligator clips. Silver paint is used on the electrode flags to protect the thin Au films from abrasion, although it is not absolutely necessary. The oscillator circuit is connected to the frequency counter, which generates either an analog (0 to 1 V) or digital (IEEE 488) signal for the computer acquisition system. The frequency counter used in our laboratories is a Philips PM 6654, capable of measuring frequencies with an accuracy of 1 part in 10^6 in as little as 6 msec. Many commercial counters would require 1 second for such a measurement. Since the rapidity of the frequency measurement is generally the factor which limits the time scale of the experiment, it is desirable to use a counter with the maximum counting rate available. Connection to the computer (an IBM PC, AT, or 80386-based clone) is made with an A/D board such as the Data-Translation DT 2801-A or using any of a number of IEEE 488 boards. The Philips PM 6654 provides for either type of connection with accessory plug-in boards.

The potentiostat is of the standard Wenking design, using a true ground for the working electrode. Potentiostatic or galvanostatic control is achieved either by using the DT 2801-A to generate staircase waveforms or by triggering an external analog ramp generator. In the absence of an external analog ramp generator, the 12 bit accuracy of the DT 2801-A forces the use of staircase voltammetry because the resolution is not sufficient to produce a staircase waveform which mimics closely enough the analog ramp required for cyclic voltammetric experiments.

With the instrument shown in Fig. 4, a typical cyclic voltammetric EQCM experiment would involve the application of the electrochemical waveform to the working electrode and (virtually) simultaneous measurement of the current flowing through the electrochemical cell and the oscillation frequency of the crystal. For large currents ($i > 1$ μA) or large frequency changes ($\Delta f > 10$ Hz), single scans are generally sufficient to achieve acceptable signal to noise ratios. Smaller signals require the use of signal averaging techniques. As always, proper attention to shielding is required for achieving the maximum signal-to-noise ratio [34]. Placing the oscillator circuit within a shielded metal enclosure has been found to be especially helpful in reducing noise levels.

Several groups have used EQCM instrumentation, which is different from that described above. For example, Ward [32] has described the use of a commercial potentiostat (PAR 173 or 273, with the working electrode at the summing point or virtual ground of an operational amplifier) connected to the EQCM oscillator in such a way that the working electrode is strapped to hard ground with the ground of the oscillator circuit. The current is then read from the voltage drop across a precision resistor in the counter electrode branch of the potentiostat. In this configuration, the commercial potentiostat is essentially converted into a Wenking style potentiostat. An advantage of such a configuration is the ready availability of such conventional electrochemical instrumentation. Bruckenstein and Shay [12] have also reported on a somewhat different version of EQCM instrumentation. Their system also allows use of conventional potentiostats, but without imposing the requirement that the working electrode be at hard ground. For all of the various EQCM derivative instruments described above stability is a critical issue. The use of a grounded working electrode would seem to be advantageous in this regard, but it is clear that other acceptable designs exist.

A variation on the apparatus shown in Fig. 4 is one in which the oscillator and frequency counter are replaced by some type of impedance analyzer or network analyzer. The use of such instruments in EQCM applications has not yet been described in the literature. However, experiments in our laboratories have shown it to be a useful complement to the more standard configuration of Fig. 4. The parameters which can be measured using such instruments and their relevance to the EQCM experiments are discussed below. Again, it is especially useful if one of the inputs to the impedance analyzer can be maintained at true ground. A Hewlett-Packard 4192A low frequency impedance analyzer is used in our laboratories for this reason.

B. Equivalent Circuit Description of the EQCM

The equivalent circuit description of the quartz crystal has been widely used to model its behavior in oscillator circuits [9,30]. Figure 5a shows the equivalent circuit for an AT-cut crystal. Figure 5b shows an equivalent representation of this circuit, which is of some use in the experimental determination of the values of the circuit elements in Figure 5a. In Figure 5a, C_O is the electrical capacitance of the quartz sandwiched between the two vapor deposited electrodes, R_1 corresponds to the dissipation of

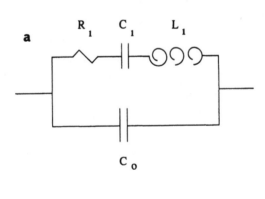

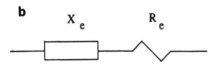

FIG. 5. (a) Equivalent circuit representing the electrical behavior of a QCM. (b) Equivalent representation of the circuit in (a).

the oscillation energy from mounting structures and from the medium surrounding the crystal (e.g., losses induced by the presence of a viscous solution), C_1 corresponds to the stored energy in the oscillation and is related to the elasticity of the crystal and the surrounding medium, and L_1 corresponds to the inertial component of the oscillation, which is related to the mass displaced during the vibration. In Fig. 5b, R_e and X_e represent the resistive and reactive components of the circuit in Fig. 5a, respectively [30]. Typical values for the components in Fig. 5a at the fundamental frequency of the crystals shown in Fig. 2 are $C_0 = 7.3 \times 10^{-12}$ F, $R_1 = 100$ Ω, $C_1 = 23 \times 10^{-15}$ F, and $L_1 = 45 \times 10^{-3}$ H. These values are those found for a crystal with both faces exposed to air. The values are different for crystals operated at their harmonic frequencies.

For a crystal with one face exposed to an aqueous solution, the value of R_1 increases to ca. 500–1500 Ω, depending on the solution, with little change in the other values, confirming the notion that R_1 can be associated with the viscous loading introduced by immersion of the crystal in the solution. The determination of R_1, which will be discussed below, thus provides a means of evaluating the extent of changes in viscous loss due to changes in the surrounding medium.

Several workers have analyzed the circuit in Fig. 5a, providing expressions for the various resonant frequencies and a few other parameters of interest [9,30]. The frequency of maximum conductance, f_s, is the resonant frequency of the motional branch of the circuit, where the conductance, G (i.e., the real part of the admittance), is a strong function of the applied frequency near the resonant point. The frequency of zero phase, f_r, occurs when the current flowing through the crystal is exactly in phase with the applied voltage. The sharpness of the resonance is measured by the Q value or quality factor of the resonator. Another parameter of interest is the capacitance ratio, r. Equations, (3) to (6) give some relevant relationships between these variables:

$$f_s = \{2\pi(L_1 C_1)^{1/2}\}^{-1} \tag{3}$$

$$Q = (2\pi f_s C_1 R_1)^{-1} = \frac{2\pi f_s L_1}{R_1} \tag{4}$$

$$\frac{f_r - f_s}{f_s} = \frac{r}{2Q^2} \tag{5}$$

$$r = \frac{C_o}{C_1} \tag{6}$$

Note that f_s and f_r are, in general, not equal, and that their separation depends on both r and Q.

Experimentally, one approach to obtaining these values is to measure the appropriate combinations of functions using an impedance analyzer, network analyzer, or bridge methods. A Hewlett-Packard 4192A impedance analyzer is used for this purpose in our laboratories. This type of instrument performs an essentially different function than does the apparatus shown in Fig. 4. An alternating voltage having a precise, synthesized frequency is applied to the crystal, usually with the capability of sweeping this frequency over a wide range. The amplitude and phase of the ac current which passes through the crystal are measured, and these quantities are used to compute various parameters of interest. Such measurements directly provide the conductance (G), the reactance (X_e), the resistance (R_e), and the phase angle (Θ), as well as several other parameters characteristic of the crystal resonator. Note that G, X_e, R_e, and Θ are functions of frequency, and in general vary strongly near the resonant point. f_r is obtained from the frequency at which $\Theta = 0°$. f_s is found from the frequency at which G is a maximum (i.e., $G = G_{max}$). R_1 is approximately equal to R_e at the frequency at which X_e is equal to zero, which is just above f_s [30]. C_o may be measured at any frequency sufficiently far from the resonant frequency of the motional branch of the circuit (e.g., 100 kHz). Thus, the measured quantities are R_1, C_o, f_r, and f_s. Equation (7), obtained by combination of Eqs. (4) to (6), gives C_1 as a function of these known quantities:

$$C_1 = (f_r - f_s)(2\pi^2 f_s^3 C_o R_1^2)^{-1}$$
(7)

Once C_1 is known, L_1 may be calculated from Eq. (3) and Q from Eq. (4) or (5).

A dimensionless quantity which is frequently referred to in discussions of quartz crystal oscillators is the figure of merit, $M = Q/r$. It has been shown [9] that when $M < 2$, then f_r will not exist, i.e., there will be no frequency at which the zero phase condition exists. For typical AT-cut crystals operating at 5 MHz, this will occur when R_1 exceeds ca. 3000 Ω. Since most oscillator circuits actually operate at a zero phase condition (which may not occur at exactly the same frequency as f_r due to influence of the components of the oscillator circuit), there will usually be cessation of oscillation when excessive viscous loss causes R_1 to be larger than this value. Values of R_1 in excess of this are rather easily obtained when thick, solvent-swollen polymer films are investigated with the EQCM. However, even under these conditions it is still possible to make meaning-ful measurements of crystal parameters (e.g., to monitor changes in G_{max} as mass loading changes) if an impedance analyzer is available.

The purpose of this section has been to briefly describe the use of impedance analysis methods for the evaluation of some of the important parameters which can be used to judge the type of behavior exhibited by a particular QCM composite resonator (i.e., the quartz, electrode, thin film, solution system). In the sections below the criteria for such judgments are discussed, with emphasis on the ability to distinguish between rigid, elastic, and viscoelastic behavior. As will be seen, such distinctions are crucial to the proper application of the EQCM to quantitative mass measurements.

C. Effects of Various Parameters on Crystal Oscillation

1. Mass-Frequency Correlations

Sauerbrey was the first to recognize the potential usefulness of AT and BT crystals as mass sensors [1]. He demonstrated the extremely sensitive nature of these piezoelectric devices towards mass changes at the surface of the QCM electrodes. He also described their differential radial mass sensitivity and correlated this with the radial distribution of the vibrational amplitude [17]. The results of his pioneering work in this area are embodied in the Sauerbrey equation (Eq. (8)), which relates the mass change per unit area at the QCM electrode surface to the observed change in oscillation frequency of the crystal:

$$
\begin{aligned}
\Delta f &= -\left(\frac{f_o}{t_q \rho_q} \right) \Delta m \\
&= -\left(\frac{2nf_o^2}{(\rho_q \mu_q)^{1/2}} \right) \Delta m \\
&= -\left(\frac{f_o^2}{N\rho_q} \right) \Delta m \\
&= -C_f \Delta m \tag{8}
\end{aligned}
$$

where Δf is the observed frequency change (Hz), f_o is the resonant frequency of the fundamental mode of the crystal (which may be significantly different than either f_s or f_r, depending on the particular oscillator circuit used), Δm is the change in mass per unit area (g cm^{-2}), $\rho_q(= 2.648$ g cm^{-3}) is the density of quartz, μ_q (= 2.947×10^{11} g cm^{-1} s^{-2}) is the shear modulus of quartz, N (= 1670 kHz mm) is the frequency constant for quartz, n is the number of the harmonic at which the crystal is being driven (i.e., 1 for the fundamental, 3 for the third harmonic, etc.), and C_f is the sensitivity factor for the crystal employed for the measurement, which depends on the thickness and, therefore, the fundamental frequency. For a 5 MHz crystal operated in its fundamental mode, C_f is 56.6 Hz μg^{-1} cm^2, so that a uniformly distributed mass increase of 1 μg cm^{-2} would result in a frequency decrease of 56.6 Hz. The sensitivity factor is a fundamental property of the QCM crystal. Thus, these mass sensors do not require calibration. This ability to calculate the mass sensitivity from first principles is a most attractive feature of these devices.

It is worth pointing out that the term Δm in Eq. (8) is an areal mass change with units of g cm^{-2}. Couching the equation in this way emphasizes the necessity of film uniformity for the quantitative evaluation of mass changes from frequency changes. It also emphasizes that, within certain limits (see Sec. II.A.1), the mass sensitivity per unit area does not depend on the electrode geometry. In other words, it is the film thickness which determines the frequency change, not its total mass.

Equation (8) shows the linear dependence of Δf on Δm, the dependence of Δf on $f_0{}^2$, and the linear increase in sensitivity which accompanies the use of higher harmonics (i.e., $\Delta f \propto n$). Thus, for a given crystal, the mass sensitivity increases linearly with the harmonic number, while for different crystals, the mass sensitivity increases as the square of the fundamental frequency. The implication of this is that sensitivity enhancements need not be made through the use of crystals with high fundamental frequencies (which are necessarily thin and fragile). Rather, such enhancements may be made by driving crystals having lower fundamental frequencies at their harmonics. This occurs at the expense of greater complexity in the oscillator circuitry.

The negative sign shows that <u>increases in mass</u> correspond to <u>decreases in frequency</u>. The rather strong functional dependence of the mass sensitivity on f_0 dictates that for precise work the deviation of the fundamental frequency from the nominal value must be taken into account. In other words, in the example above C_f would need to be modified appropriately if f_0 were much different from 5 MHz (e.g., 4.9 rather than 5 MHz).

Equation (8) is indirectly based on Eq. (1) in that the incremental change in mass from the foreign film is treated as though it were really an extension of the thickness of the underlying quartz disk. Thus, the material properties of the foreign film are not considered explicitly in Eq. (8). The foreign film is considered to be so thin that it exists entirely at the antinode of the vibration, so that it does not experience any shear forces. When this implicit assumption does not hold true, then the material properties of the foreign film must be taken into account.

Sauerbrey's early work has been extensively discussed and extended (see Ref. 2 and references therein). Significant improvements in later treatments have centered around the explicit incorporation of the elasticity (i.e., the shear modulus) of the deposit [35,36] and the extension of the measurement to higher harmonics [37]. Lu and Lewis [36] gave an especially simple equation for the dependence of Δf on Δm:

$$\tan \frac{\pi f_c}{f_0} = - \frac{z_f}{z_q} \tan \frac{\pi f_c}{f_f} \tag{9}$$

where f_c is the resonant frequency of the composite resonator formed from the crystal and the film(s) present at the surface, f_f can be thought of as the resonant frequency of the free-standing foreign film, and z_f and z_q are the acoustic impedances of the film and quartz, respectively. Equation (8) can be shown to be an approximation of Eq. (9) in which the tangent function is expanded in a power series, retaining only the first term. Some equations which interrelate these parameters are:

$$\Delta f = f_c - f_o \tag{10}$$

$$f_f = \frac{v_f}{2t_f} = \frac{v_f \rho_f}{2 \Delta m} \tag{11}$$

$$v_f = \left(\frac{\mu_f}{\rho_f} \right)^{1/2} \tag{12}$$

$$v_q = \left(\frac{\mu_q}{\rho_q} \right)^{1/2} \tag{13}$$

$$z_q = \rho_q v_q = (\rho_q \mu_q)^{1/2} \tag{14}$$

$$z_f = \rho_f v_f = (\rho_f \mu_f)^{1/2} \tag{15}$$

where $t_f (= \Delta m / \rho_f)$ is the film thickness, μ_q (= 2.947×10^{11} g cm^{-1} s^{-2}) and μ_f are the shear moduli of quartz and film, respectively, ρ_f is the density of the film, and v_q and v_f are the velocities of the acoustic waves in the quartz and film, respectively. This analysis of frequency changes using the acoustic impedances of the quartz and film is usually called the Z-match method.

The choice of whether to use Eq. (8) or (9) for determining the mass change for deposition or dissolution processes of rigid films at electrode surfaces depends on the thickness of the deposit to be investigated. (Considerations pertaining to nonrigid films will be discussed below.) When the mass loading from the deposit causes a change in the resonant frequency of less than 2% of f_o, then Eq. (8) may be used. Equation (9) is accurate for frequency changes of up to ca. 40% of f_o [36]. The degree to which Eq. (8) deviates from the more accurate Eq. (9)

depends on the ratio of acoustic impedances for quartz and the deposit.

Use of the Z-match method requires that the shear modulus of the deposit be known. While this will not be a problem for many cases (e.g., some metal or metal oxide bulk depositions), in other instances the properties of the deposit may be completely unknown or may be very different from those of the bulk material. It has been asserted that the density of solid materials varies more strongly than does the shear modulus from one material to another [37]. Thus, an approximation for the z of a deposit may be obtained (if its density is known) by assuming that its shear modulus is the same as that of quartz (i.e., $z_{f,approx} = (\rho_f \mu_q)^{1/2}$). However, this provides only a very rough approximation of z_f, and still requires that ρ_f be known.

An alternative method [37] makes use of the fact that the response of the crystal to varying acoustic impedance of the foreign film is different at the fundamental and harmonic frequencies. Measurement of the change in resonant frequency at more than one frequency (usually the fundamental and third harmonic) thus provides a method for obtaining z_f, allowing the use of Eq. (9) for the accurate calculation of mass changes. This method will be useful only when the relative frequency change is large (greater than ca. 10%), because this is the condition required for significant deviation from Eq. (8). Experimentally, this method requires the use of an oscillator that can be easily switched between operation at the fundamental and third harmonic frequencies or the use of an impedance analyzer which can operate at a frequency high enough to measure the parameters of interest at the third harmonic. Details of the method have been described [37].

Even though the influence of the deposit elasticity on the frequency change can be easily dealt with, it will frequently be true that Eq. (8) provides sufficient accuracy, because the film

thickness (and, therefore, the relative frequency change) will
be small. However, this situation holds only for rigid deposits.
Films which are behaving viscoelastically, such as some organic
polymer films with large thickness or viscosity, will exhibit significant deviations from both Eqs. (8) and (9). The influence
of such effects is discussed below.

2. Effect of the Solution on Oscillation

When a planar surface that is oscillating in a shear mode is immersed into any medium, a shear wave propagates normally away
from the surface into the medium. Because of the energy dissipation caused by the viscous response of liquids, the shear
wave is damped exponentially as it travels through the liquid
away from the surface. This situation, as applied to oscillating
quartz AT shear wave transducers, was first treated by Glassford [38] and later by Kanazawa and Gordon [39,40]. The
equation for the shear wave velocity as a function of distance
from the surface of the oscillating crystal was given [39]:

$$v(z,t) = v_o \exp(-k_1 z) \cos(k_1 z - 2\pi f_o t) \qquad (16)$$

where $v(z,t)$ is the shear velocity (parallel to the surface),
which is a function of distance (z) from the surface (which is
at $z = 0$) and time (t), v_o is the velocity of the crystal surface,
k_1 is the propagation constant, and the other quantities have
been defined. The propagation constant k_1 is given by:

$$k_1 = \left(\frac{\pi f_o \rho_1}{\eta_1} \right)^{1/2} \qquad (17)$$

where ρ_1 and η_1 are the density and viscosity of the liquid,
respectively. Equation (16) shows the damped sinusoidal nature
of the shear wave, and Eq. (17) shows the dependence of the
damping on the density and viscosity of the medium. Figure 6
[40] shows the shear velocity profiles in the fluid at three

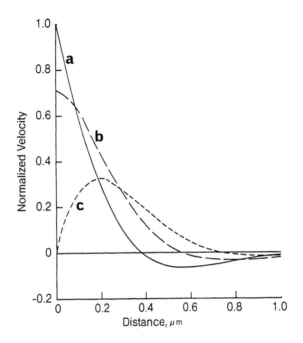

FIG. 6. Shear velocity profiles in a fluid adjacent to a QCM at three different times: (a) peak surface velocity, (b) intermediate surface velocity, (c) zero surface velocity. (From Ref. 40, with permission.)

different times: at peak surface velocity, at intermediate surface velocity, and at zero surface velocity.

The reciprocoal of the propagation constant is the decay length of the shear wave. For pure water at 20°C, with ρ_l = 0.9982 g cm^{-3} and η_l = 1.002 × 10^{-2} g cm^{-1} s^{-1}, this length is ca. 250 nm. Thus, only the first μm or so of the solution is perturbed by the shear displacement. Since these displacements are parallel to the surface and rather small (10–100 nm, see above), there should be negligible stirring effects, as has been found by workers in our laboratories and others.

Kanazawa's treatment of the influence of the solution properties on the crystal permits the prediction of the change in resonant frequency which accompanies immersion of the crystal into
a viscous medium:

$$\Delta f = -f_o^{3/2} \left(\frac{\rho_1 \eta_1}{\pi \rho_q \mu_q} \right)^{1/2} \tag{18}$$

where the decrease in resonant frequency is seen to be proportional to $(\rho_1 \eta_1)^{1/2}$. This equation predicts a decrease in f_o of
ca. 700 Hz on transfer from vacuum to pure water at 20°C. Due
to experimental difficulties associated with mounting stresses, it
has been more common to measure frequency changes for different
solutions relative to pure water. Very good agreement with Eq.
(18) for such relative changes has been obtained for aqueous
solutions of glucose, sucrose, and ethanol using this type of
differential measurement [39,40]. Linearity to only the solution
viscosity has also been demonstrated [12], indicating that density effects can be neglected under certain conditions, but this
will not be true in general.

Based on the results of an analysis similar to Kanazawa's,
Hager has proposed that the QCM can be used in the evaluation
of fluid properties, specifically $(\rho_1 \eta_1)^{1/2}$ [41]. A significant
result of this contribution was the demonstration that the Gaussian velocity distribution of fluid at the crystal surface can be
replaced by a surface average velocity, making the analysis of
the problem considerably more tractable. He also proposed the
use of multiple crystal arrays for the separate determination of
ρ_1 and η_1 from the frequency shifts caused by exposure of the
crystals to a fluid.

In a recent contribution, the equivalent circuit description
of the QCM has been used to derive an equation relating R_1 to
$(\rho_1 \eta_1)^{1/2}$ [42]. Impedance analysis techniques were used to
arrive at a value of R_1, and this was plotted versus $(\rho_1 \eta_1)^{1/2}$

for water/ethanol and water/glycerol mixtures. Good linearity
was observed except for the case of crystals which had both
electrodes immersed in solutions with high proportions of water.
This deviation was rather speculatively attributed to unusual
dielectric properties of water or to unknown electrical effects.
However, the important finding was the linearity of R_1 to
$(\rho_1 \eta_1)^{1/2}$. Thus, R_1 serves as a good measure of viscous loading by the medium at the crystal surface.

When the EQCM resonator is loaded by viscous coupling to a
liquid, its conductance spectrum exhibits predictable changes.
Recalling Eq. (4), which shows the reciprocal dependence of the
crystal Q on the value of R_1, it is predicted that the increase
in viscous loss should appear as an increase in R_1 and, therefore, as a decrease in Q. The methods described in Sec. II.B
may be used to measure the changes in R_1.

A more intuitive, graphical way of measuring the changes in
viscous loading is provided by plots of conductance (the real
part of the admittance) versus frequency, which can be conveniently obtained using impedance or network analyzers. Figure 7 shows such a plot for a 5 MHz crystal, mounted in the
o-ring mounting as in Fig. 4, both in air (curve A) and in water
(curve B) at room temperature. Curve A shows the sharpness
of the resonance when viscous loading is absent, while curve B
(note the difference in vertical scales for the two curves) shows
the dramatic increase in the width of the resonance and decrease
in the maximum conductance which accompany immersion in the
liquid. The size of the frequency shift, ca. 800 Hz, is in reasonable agreement with the prediction from Eq. (18) (700 Hz).
A measure of the change in Q is obtained graphically by comparison of the widths at half height for these conductance plots.
In this way, the Q is found to decrease by a factor of ca. 15
upon immersion of the crystal in water. As will be discussed

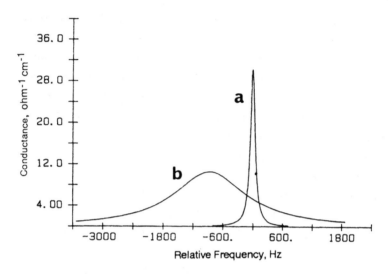

FIG. 7. Plots of conductance versus relative frequency for a 5 MHz AT-cut QCM of the type shown in Figure 2 in (a) air, (b) water. The scale for curve b has been increased by 5x.

below, the changes in Q induced by changes in the viscoelastic properties of the deposit may also be monitored by such impedance techniques.

The dependence of f_O on the properties of the liquid at the crystal surface has important consequences for experiments in which transfer from one solution to another is needed (e.g., in measuring changes in swelling for polymer films in different solutions). One must take special care to account for the different values of f_O which obtain in the different solutions by measuring these offsets with bare crystals. Even then, subtle changes in the boundary conditions can render such measurements suspect. For example, the exact position of the slip plane, which defines the position in the solution which moves with the same amplitude and phase as the underlying crystal surface, may not remain constant upon transfer from one solution to another. Also, the

degree to which solvent trapping in pores at the surface occurs
may change for a bare crystal and a coated crystal, so that
simple subtraction of the offset may not always be possible.
Other possible sources of discrepancy also exist. On the other
hand, it has been amply demonstrated that when experiments
involve only relative frequency changes which are measured in
a given solution, the offset caused by the viscous loading has
negligible effect on the accuracy of Eq. (8) for the determination
of small mass changes.

3. Temperature Dependence of Oscillation

The intrinsic dependence of the resonant frequency of a quartz
crystal on temperature (T) is caused by changes in ρ_q and μ_q
with T. These effects are well known and generally quite small,
being around 1 Hz per °C depending on the particular cut of
the crystal [2]. In fact, the AT-cut is popular partially because it is cut specifically to give a near-zero T coefficient at
room temperature. Thus, the intrinsic T dependence of the
crystals is practically negligible in EQCM applications. However,
much larger changes in f_O with T occur when crystals are immersed in solution due to the coupling of the acoustic shear
wave into the solution as described above. Equation (18) shows
that f_O is proportional to $(\rho_l \eta_l)^{1/2}$ and allows calculation of the
effect of changes in T given the density and viscosity as functions of T. For example, the values of ρ_l at 20 and 25°C are
0.99823 and 0.99707 g cm^{-3}, respectively, and the values of η_l
at 20 and 25°C are 1.002 × 10^{-2} and 0.8904 × 10^{-2} g cm^{-1} s^1,
respectively. Use of these values in Eq. (18) to calculate the
expected change in f_O gives 41 Hz. This frequency change is
larger than those for monolayer adsorption/desorption processes
and not negligible with respect to those for solvent or ionic
transport in thin films. Thus, in experiments in which the frequency is to be monitored at length, the temperature must be
controlled to at least 0.1°C, preferably to better than 0.02°C.

This is possible with commercially available temperature control baths and jacketed cells. On the other hand, measurement of short-term, relative frequency changes (which might occur during a cyclic voltammetric experiment, for example) generally do not require such careful control of the temperature because the drift in T, which can occur during a scan lasting only a few seconds, is negligibly small.

4. Viscoelastic Deposits

An attractive use of the EQCM is to study the mass transport processes which occur during redox events in thin polymer films. Due to their large molecular weights and the resulting chain entanglement, to cross-linking (whether chemical or physical), and to solvent-induced swelling, these systems frequently exhibit viscoelastic behavior. It is extremely important to determine whether the viscoelastic nature of the deposit is influencing the resonant frequency of the crystal oscillator, because the linear mass-frequency correlation discussed above will, in general, not hold under such conditions.

An excellent source for general information on the viscoelastic properties of organic polymers is the classic book by Ferry [43]. In general, the viscosity and shear moduli of polymeric solids, gels, and solutions are functions of the frequency at which they are measured [43]. The shear modulus (μ), a measure of the stiffness of the material, increases with frequency, while the viscosity (η) tends to decrease with frequency. At MHz frequencies, polymeric systems will tend to have values for μ lower than that for quartz due to their lower stiffness, while having values for η higher than that for quartz due to the possibility for viscous loss from translational motion of the chains relative to one another. Since rigid layer behavior will tend to prevail when μ is large and η is small, the higher the frequency, the greater the chance of observing rigid layer behavior (i.e., the behavior of elastic, as opposed to viscous, deposits). Also, since μ increases and η decreases below the glass transition temperature (T_g), a knowledge of T_g for the film material will be useful in gauging whether or not viscoelasticity in the deposit is influencing the resonant frequency of the EQCM. Films with T_gs far above the temperature of the experiment will tend to behave more rigidly than those with lower T_gs. Also, η will tend to increase for films when they are swollen by solvent. Thus, the conditions under which the measurements are made (high frequency, above or below T_g, swelling solvent or not, etc.) influence the degree to which rigid layer behavior prevails, and therefore, the extent to which mass changes can be inferred directly from Δf using the Sauerbrey equation or the Z-match method.

The quantitative connection between the material properties of the film (i.e., μ and η) and the behavior of the EQCM resonator has yet to be established for the important case of a thin film on the EQCM in a semi-infinite liquid, the configuration of most polymer-modified electrodes or thin film sensors based on the QCM. The problem is complex because the shear wave exists simultaneously in the quartz crystal, the thin film, and the adjacent solution, so reflection of the shear wave at several interfaces must be taken into account. It would be most worthwhile to solve this problem, because then μ and η for the film could be calculated from the information provided by an impedance analysis of the film. This would allow for correlation of the electrochemical behavior of the film (i.e., charge transport rates, permeability, etc.) with the material properties of the film, clearly a worthy goal. Reed, Kanazawa, and Kaufman have recently solved the problem of a viscoelastic film of arbitrary thickness atop the QCM resonator, taking the piezoelectricity of the quartz explicitly into account [44]. This contribution will be a valuable one, and we hope will lead the way

for an analysis which takes into account the shear wave propagation into the liquid adjacent to the film.

As was pointed out earlier, it is essential to determine
whether or not the viscoelastic properties of the film influence
the measurement. For example, if mass measurements are being
made of a film deposition process, it is possible that as film
thickness increases, the mass sensitivity may change due to excessive viscosity of the deposit (e.g., this would be a problem
common to highly solvent swollen polymer films). There are
several ways of checking for such effects.

One method is to verify that changes in deposit thickness or
surface coverage scale linearly with changes in Δf. This requires that some parameter be found which provides an independent measure of thickness or coverage, such as electrochemical
charge, optical absorbance, ellipsometric functions, etc. Then,
to the extent that the changes in these quantities scale with
changes in the observed Δfs, a good argument can be made that
the measured Δfs are a true representation of the mass changes
occurring at the surface. However, many effects could obscure
such comparisons, an example of which would be varying porosity or density of the deposit as a function of thickness (an effect which is rather common in electrodeposited oxide and conducting polymer films).

A more straightforward and less ambiguous method relies on
the dependence of the crystal Q on the viscosity of the deposit.
Thus, measurements of Q during a film deposition will reveal the
extent of changes in the deposit viscosity which can signal the
potential for deviations from rigid layer behavior. In particular,
as Q decreases due to increasing deposit viscosity, the width of
the G versus f plot will increase above the value which arises
from immersion of the bare crystal in solution. It is not yet
clear how large this increase must be to generate significant deviation from Eqs. (8) or (9), but the presence of detectable

increases in Q (e.g., >10%) should certainly give rise to the suspicion that such deviations might exist.

The range of Q values which can be observed is quite large. For example, for crystals in air, Q can range from 10^4 to 10^6, depending on crystal finish and the details of electrode geometry. Immersion into solution usually decreases Q to values on the order of 10^3. For the case of a crystal loaded by an extremely viscous film (such as a highly plasticized or solvent swollen) polymer, Q can decrease to values near 10^2. However, in determining the influence of viscous loading for a given crystal/film combination, the important determination is whether the presence of the film causes a change in Q from the value observed for the bare crystal in the same solution.

5. Deposit Roughness and Porosity

As mentioned above in the context of crystal surface finishes, roughness can cause large apparent mass loadings due to the liquid which is trapped within pores at the crystal surface. This can occur not only from roughness of the quartz crystal surface, but also from roughness which might exist in the deposit. Schumacher et al. [22] showed that such roughness in gold oxide deposits can cause very large discrepancies between the expected Δf and the observed value. They attempted to fit the discrepancies by modeling the surface as having hemispherical undulations which trapped solution, and further assuming that this trapped solution behaved as a mass rigidly attached to the surface [22]. Although some disagreement between this model and the experiment data was observed, general trends suggest that the qualitative features of the model are sound. In a later contribution, Schumacher et al. [23] gave other experimental examples of the appearance of such roughness effects for Ag and Cu surfaces following oxidation. They also showed that, for the case of Ag, prolonged application of sufficiently negative potentials caused the surface to become relatively smooth again, as judged by both the increase in EQCM frequency due to the loss of trapped solution and by differential capacitance measurements. Again, however, quantitative predictions of the expected frequency changes from the roughness based on SEM observation of the roughened surfaces were not in agreement with the experimentally observed Δfs.

Experiments recently done in our laboratories have also indicated that deposit morphology can influence the EQCM frequency response. The deposition and dissolution mechanisms of electrochromic films of diheptylviologen bromide (DHVBr) were studied using the EQCM [45]. In this work, the dissolution at fast scan rates of the DHVBr films was shown to proceed by a pitting mechanism, with the pits eventually growing together over time so that the film disintegrated only late in the scan. This situation is shown schematically in Fig. 8. To the extent that the solvent trapped within the pores has a density similar to that of the film, the frequency should only increase late in the scan, when the disintegration occurs. This is because the mass lost by film dissolution is replaced by the solution trapped within

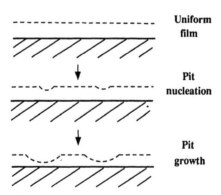

FIG. 8. Schematic depiction of pit nucleation and growth during thin film dissolution.

the pores. The observation of this behavior in the EQCM experiments prompted the proffering of a model for dissolution involving pit nucleation and allowed for a (rather speculative) assignment of pit nucleation as being progressive rather than instananeous, based on the scan rate dependence of the frequency changes. This study serves as an example of the use of indirect EQCM information regarding surface roughness to help elucidate mechanistic details about the behavior of thin films.

In the measurement of monolayer mass changes, the magnitude of the observed frequency change is frequently so small that meaningful mass changes cannot be made using the usual experimental conditions. One way around this is to purposefully roughen the surface to increase the surface area of the electrode. Since the sensitivity to mass changes increases with the true surface area, this strategy makes it possible to increase the detection limit for mass changes linearly with the increase in surface area. Preliminary experiments have indicated the feasibility of this approach for sensitivity enhancements of approximately a factor of two. Larger increases should be possible. A disadvantage of such an approach is that the measurement of absolute frequency changes cannot be made due to the relatively unpredicatable effects of solvent trapping within the pores on the roughened surface.

Another possible source of frequency changes for thick films is porosity of the deposit. If a film is porous, then significant amounts of mass may be trapped (as solution) within the pores of the film (i.e., in the void volume of the film). In this case the frequency decrease for deposition of the film will result not only from the mass of the film itself, but also from the mass of solution trapped within the pores. In the limit of microscopic pores, this model reduces to one of film swelling by the solvent or solution, so that all of the effects discussed above regarding film swelling hold here. To the extent that the film behaves elastically both with and without the trapped solvent, it is possible to calculate the amount of trapped solvent (and, therefore, the pore volume). This is done from the difference in the oscillation frequency of the crystal coated with the (dry) film in air and with the hydrated crystal in solution. For this calculation, the offset due to solution viscous loading must be corrected for and, therefore, must be the same for the bare and coated crystal. For highly porous films, roughness effects will be important, so that these types of calculations will only be feasible for very thick films for which the contribution from surface effects (e.g., roughness) will be small in comparison to the total frequency change.

D.   Correlations Between Frequency Changes and Electrochemical Parameters

In correlating EQCM frequency changes with electrochemical data, it is useful to use different methods of data presentation. For convenience, the case of application of the EQCM to the study of a simple deposition process will be considered. An example would be the electrodeposition of Ag. The two important electrochemical parameters to compare to Δf are the charge (Q) and the current (i). The charge is an integral measure of the total number of electrons delivered at the interface during the process (reduction, in this case). To the extent that each electron supplied results in the deposition of one atom of Ag, there should be direct proportionality between Q and Δf. This is shown by Eq. (19):

$$\Delta f = \frac{10^6 \ MW \ C_f Q}{nF} \tag{19}$$

where n is the number of electrons transferred to induce deposition, F is the Faraday constant, MW is the apparent molar mass of the depositing species, C_f is the sensitivity factor for the crystal employed (see Eq. (8)), and the factor of 10^6 provides for the unit conversion from μg in C_f to g in MW. A plot of Δf versus Q will give the apparent mass per electron of the deposited species, when n is taken into account. The corresponding equation for the relationship between i and Δf is:

$$i = \frac{\{d(\Delta f)/dE\}\ (10^{-6}\ n\ v\ F)}{MW\ C_f} \tag{20}$$

in which $d(\Delta f)/dE$ is the derivative of Δf with respect to potential (for a cyclic voltammetric experiment), v is the scan rate, the factor of 10^{-6} provides for unit conversion described above, and the other parameters have been described above. Equation (19) is useful for comparing total changes in oscillation frequency with electrochemical charges, while Eq. (20) is especially useful for the detection of subtle relationships between Δf and i. These equations will be used in the sections below to provide for detailed comparison between frequency and electrochemical parameters for various systems. Of course, electrochemical experiments other than cyclic voltammetry may be used in conjunction with the EQCM. The appropriate equations for these may be easily derived, keeping in mind the proportionality between Q and Δf.

III. MONOLAYER SYSTEMS
A. Introduction

EQCM studies of the electrosorption of monolayer systems are an extremely challenging application of the technique due to the very small mass changes which occur. In addition to the electrosorption event, several other processes may contribute to the observation of frequency changes, such as adsorption/desorption of solvent or supporting electrolyte ions during the electrosorption, changes in the position of the slip plane (i.e., the plane which defines the first layer of the adjacent medium which is

not rigidly attached to the electrode surface) due to the electrosorption, and changes in the viscosity and/or density of the
double layer induced by changes in composition as a function of
applied potential. These effects have not yet been unambiguously
observed, but few systems have been quantitatively studied as
of yet.

One feature of monolayer systems which makes them somewhat more tractable is that, since the adsorbate layer is so thin
compared to the wavelength of the shear wave in the resonator,
the added mass from the electrosorbed layer should exist entirely
at the antinode of the shear wave. Therefore, the adsorbate
layer should experience no significant shear deformation, and its
viscoelastic properties should have no influence on the measurement.

The studies described in this section demonstrate the detection of submonolayer mass changes during Faradaic processes at
electrode surfaces. In most, mass changes are directly correlated
with charge to gain insight into the compositional changes accompanying the interfacial charge transfer event. For example, it
is possible to determine the electrosorption valency from such
measurements. In other cases, deviations in the predicted correlations are speculatively attributed to the influence of solvent
and/or ion association with the monolayer. Much remains to be
done to elucidate how interfacial interactions contribute to the
observed frequency change.

B. Electrosorption of Oxides and Halides on Au

The first application of the EQCM to the in situ measurement of
mass changes from monolayer deposition or dissolution processes
was by Bruckenstein and Shay [46], who studied the formation
of the adsorbed oxygen monolayer at Au electrodes. They were
able to observe frequency changes for this process as small as
ca. 15 Hz with a signal-to-noise ratio of better than 50. A 10 MHz

crystal operating in its fundamental mode was used in this study, for which the mass sensitivity is ca. 1×10^{-9} g/Hz. Frequency decreases were observed for formation of the adsorbed monolayer, and these frequency changes were reversible in the sense that the original frequency was reattained when the adsorbed layer was removed by reduction. The agreement between the mass change predicted from the charge measurements and that inferred from the frequency change was better than 10%. Hysteresis between the time of passage of electrochemical charge and the observed mass change (i.e., 40% of the charge for a monolayer was passed before any mass change occurred) was used to support a postulated place exchange mechanism for O electrosorption (in which O electrosorption is followed by exchange of Au and O to give a "buried" monolayer of O and an exposed monolayer of Au), which was consistent with a number of previously proposed mechanisms. This first application of the EQCM to the study of a monolayer system revealed the power of the technique for making detailed mechanistic proposals based on precise mass measurements during redox events at electrodes.

These workers also observed gradual frequency changes which occurred in the double layer region of the voltammetric scans. These frequency changes were speculatively attributed to changes in the structure of the double layer (e.g., adsorption of supporting electrolyte anions at potentials positive of the point of zero charge).

Schumacher et al. [22] also used the EQCM to study the formation of electrosorbed oxide layers on Au electrodes. These workers observed mass gain coincident with formation of the oxide layer, however, their conditions promoted the roughening of the surface, in contrast to those of Bruckenstein and Shay [46]. As discussed in Sec. II, this roughening caused larger than expected frequency decreases for oxide formation, especially in basic and neutral media. This result was confirmed in a later

study by Stöckel and Schumacher [27], who also observed that
frequency decreases in aqueous sulfuric acid are smaller than
expected. They postulated that place exchange only occurs in
neutral and basic media, with monolayer discharge in acid perhaps accompanied only by deprotonation of adsorbed water.
While clearly of a speculative nature, these results point to the
type of information available from the EQCM.

The electrosorption of bromide and iodide on Au electrodes
was studied by Deakin et al. [47], who showed that full monolayer coverages were attained at sufficiently positive potentials.
This conclusion followed from the evaluation of the mass changes
which occurred during the electrosorption process and the geometric constraints imposed by the sizes of the halide ions. The
slopes of plots of the charge for electrosorption versus surface
coverage (obtained from the EQCM frequency change) were shown
to provide the electrosorption valency, γ (the ratio of the number of moles of charge passed to the number of moles of adsorbate deposited). Values for γ obtained in this way were reported as 1.01 ± 0.05 for iodide and 0.30 ± 0.03 for bromide,
both in excellent agreement with previously reported values from
other groups. This was one of the first reports to demonstrate
the use of signal averaging methods for the acquisition of EQCM
data and clearly demonstrated the extension of the method to
detailed, submonolayer mass/charge correlations.

C. Underpotential Deposition of Metals

The first application of the QCM to UPD processes was by
Bruckenstein and Swathirajan [48]. These authors made ex
situ measurements of mass changes due to UPD of Pb and Ag
on Au electrodes by removing the QCM crystal from the cell
following the deposition and measuring Δf for the dried crystal
in air. They were able to determine UPD coverage with good
accuracy, as compared to parallel rotating disk and rotating

ring disk electrode measurements. The reported accuracy was
5–10%. A 10 MHz crystal with a sensitivity factor of ca. 0.227
Hz/ng cm^2 was used in this study.

The first in situ application of the EQCM to UPD processes
was the determination of the electrosorption valency for Pb UPD
on Au by Melroy et al. [24]. A value for γ of 2.0 was obtained
by comparison of the total charge and the total mass gain for a
scan over the UPD region. The accuracy for this determination
was ca. 5%, and the value for γ was in good agreement with
those previously reported. This study employed a 5 MHz crystal
driven at its third harmonic frequency (15 MHz) through the use
of a tuned LC element in the feedback loop of the oscillator circuit, as described above [24]. The use of the third harmonic
provided a threefold larger mass sensitivity than would have
been obtained at the fundamental frequency, as shown by Eq.
(8). In this case, the mass sensitivity was 0.170 Hz/ng cm^2.

Deakin and Melroy [28] reported on a very detailed study
of the UPD of Pb, Bi, Cu, and Cd on Au. They used a com-
puterized apparatus which provided for facile manipulation of
the frequency and electrochemical data. This allowed them to
easily construct plots of charge versus mass, the slopes of
which are equal to $F\gamma/MW$, providing a simple determination of
γ. This analysis follows along the same lines as that used to
arrive at Eq. (19). Further, they demonstrated the use of the
derivative representation of Δf (see Eq. (20)) for the direct
comparison of the rate of mass change with the electrochemical
current. The data in Fig. 9 serve as an example of the type
of information which can be obtained from such an analysis. The
dashed line is the cyclic voltammetric current for a scan through
the UPD region and into the region for bulk deposition of Bi.
The solid line is generated from Δf using Eq. (20) and assuming
that γ = 3. It is immediately evident that the voltammetric
peaks which occur near 0.15 V are accompanied by mass changes

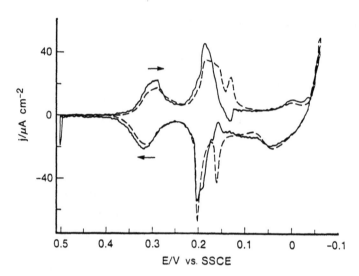

FIG. 9. UPD of Bi (1.0 mM in 0.1 M HClO$_4$) on Au at 10 mV/sec, average of 30 scans, background corrected. The current response of the EQCM Au electrode (dashed) and the derivative representation of d(Δf)/dE versus E from Eq. (20) (solid), assuming γ = 3. (From Ref. 28, with permission.)

which are much smaller than and in a direction opposite to expectations. The authors presented two possible explanations for this effect. Referencing to the negative scan, first, the current may be from further discharge of adsorbed (but only partially reduced) Bi, with a consequent desorption of weakly adsorbed anions. Second, the reduction may induce some change in electrode morphology which influences the coupling of the crystal to the solution, thereby changing the frequency in such a way as to obscure any mass flux which may accompany this reduction. A study of the anion dependence of this effect would be an obvious test of the first hypothesis, which seems the most reasonable of the two. At any rate, this study showed that use of the derivative representation of Δf clearly allows one to observe very subtle relationships between current and mass flux.

D. Adsorption/Desorption of Surfactant Molecules

Work in our laboratories has shown that mass changes associated with electrochemically induced changes in surface coverage of redox surfactants can be monitored using the EQCM [49,50]. These redox surfactants are derivatives of (dimethylamino)methyl-ferrocene ($(CH_3)_2NCH_2Fc$ or DMAFc, where Fc is the ferrocene group), which have been quaternized at the amino nitrogen using 1-bromoalkanes having chain lengths ranging from 1 to 18 carbons. These are referred to by the number of carbons in the chain, so that the 18-carbon derivative would be called C18, the 12-carbon derivative, C12, and so on. This class of molecules was first reported by Saji et al. [51], who showed that these surfactant derivatives form micelles when the ferrocene group is present in its Fe(II) form (when the molecule is a cation), while oxidation to the Fe(III) form (the dication) results in disruption of the micelles. This behavior suggested that the tendency toward adsorption for these surfactants could also be electro-chemically modulated, a notion which has since been verified [49,50].

Figure 10 shows results that are representative of this be-havior. Curve A shows the cyclic voltammogram observed for a 22 µM solution of C12. At this low concentration, the contribu-tion to the current from diffusion is negligible, and the only response is from those molecules of C12 which are adsorbed at the Au electrode surface. Thus, a symmetric surface wave centered near 0.48 V is seen. Curve B shows the EQCM fre-quency response obtained simultaneously with the CV data. These data clearly show a loss of mass (frequency increase) coincident with the redox process at 0.48 V. Most of the mass is regained during the return scan, but over a slower time frame due to the slow delivery of C12 to the surface from this extremely dilute solution. For this same reason, the cathodic wave in the CV is smaller than the anodic wave.

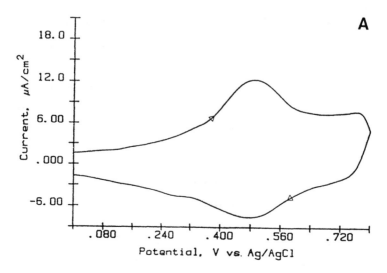

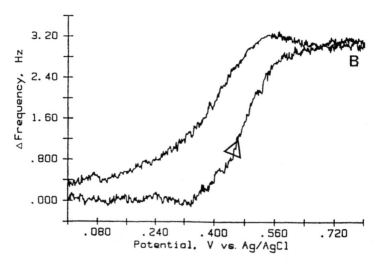

FIG. 10. (A) CV scan for 22 μM C12. (B) EQCM frequency
response for scan in (A). Scan rate—50 mV/sec. Supporting
electrolyte—1.0 M H_3PO_4. The data have been digitally smoothed
using a Fourier filter and background corrected. (From Ref.
50, with permission.)

The implication of these data is that the oxidation of C12 to its dicationic form results in its desorption, and a corresponding mass loss. Direct comparison of the mass loss with the charge required to electrolyze the adsorbed layer indicates that MW (see Eq. (19)) is, within experimental error, equal to the molar mass of the surfactant derivative, excluding the mass of any counterion or solvent which may associated with it. In other words, the frequency increase can be accounted for entirely from the electrochemically induced desorption of the surfactant itself.

The value of Δf of 3.2 Hz corresponds to a mass loss of 53 ng cm^{-2} or a loss of 1.3×10^{-10} mole cm^{-2}. This is roughly half of a monolayer, based on a head group limited molecular are of 50 square angstroms (estimated using molecular models). It is clear from the signal-to-noise ratio in the figure that the detection limit is at least as small as 5% of a monolayer. These data were obtained with a 5 MHz crystal operated at its fundamental frequency and are the average of 10 scans.

This very close agreement between predicted mass loss and charge consumed during desorption does not hold for other chain lengths. For example, for the C10 derivative, the mass loss is larger than predicted when the surfactant is present at submonolayer coverages [50]. For the C14 derivative, the mass loss is also larger than predicted, but only after complete monolayer coverage has been achieved. (These molecules exhibit Langmuirian adsorption isotherms with saturation at one monolayer [50].) These effects have been tentatively attributed to (1) solvent incorporation into the monolayer for C10 at low coverages, so that the mass for desorption includes a contribution from loss of this trapped solvent, and (2) association of counterions with the high charge density C14 monolayer at saturation coverage so the mass for desorption includes a contribution from these tightly bound counterions. Such effects are not without precedent [50]. While of a highly speculative nature, these

proposals from preliminary studies of monolayer adsorption/
desorption processes are indicative of the wealth of unique information which is, in principle, available from the EQCM.

Another benefit of using the EQCM to study adsorption/
desorption processes of such molecules is the ability to observe
these processes in the presence of large excesses of solution
phase species. Figure 11 illustrates this point [49]. The voltammogram (curve a) shows clearly the wave for the C12 derivative in solution (0.5 mM). The surface wave for the adsorbed
redox couple is not readily apparent due to the relatively large
current for the solution phase couple. However, the EQCM frequency response shown in curve b clearly shows the mass loss,
which is a consequence of the oxidation of the surface species.
Thus, the adsorption/desorption behavior of the adsorbed couple
may be inferred from the mass flux, even under conditions in
which the electrochemical response is dominated by the solution
phase couple.

The longer chain derivatives (e.g., C16 and C18) are more
strongly adsorbed at the Au electrode surface. Studies of the
mass changes which occur during redox cycling of these tightly
bound monolayers give more convincing evidence that ion association can be probed using the EQCM. Figure 12a shows the voltammetry of a monolayer of adsorbed C16 at 50 mV/s. The morphology
of the waves is similar to those described above and essentially in-
variant with scan rate. Figure 12b shows the corresponding
EQCM data, but at three different scan rates. At the lowest
scan rate, a smooth frequency increase like that in Fig. 10 is
observed, characteristic of mass loss during oxidation as described above. However, at higher scan rates, the frequency
shows a pronounced decrease prior to this increase. This is
attributed to a desorption rate for C16 which is slow enough
on this time scale so that charge compensating anion insertion
into the monolayer occurs prior to the ultimate desorption event

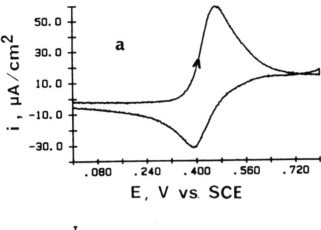

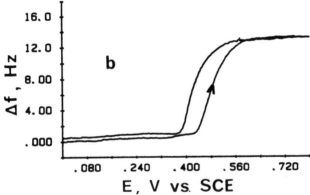

FIG. 11. (a) CV scan for 0.5 mM C12. (b) EQCM frequency response for scan in (a). Scan rate—50 mV/sec. Supporting electrolyte—0.2 M Li_2SO_4, adjusted to pH 3 with 0.2 M H_2SO_4. The CV has been treated as in Fig. 10. (From Ref. 49, with permission.)

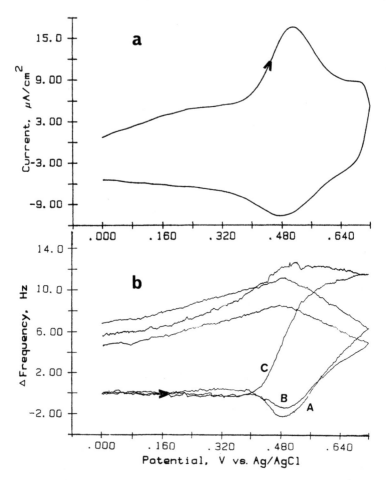

FIG. 12.   (a) CV scan of 15 μM C16.   Scan rate—50 mV/sec.
(b) EQCM frequency response for the scan in (a).   Scan rates—
(A) 250 mV/sec.   (B) 50 mV/sec.   (C) 10 mV/sec.   Supporting
electrolyte—1.0 M H_2SO_4.

[52]. These rather speculative interpretations suggest that transient mass transport phenomena can also be observed with the EQCM, given the right conditions.

The objective of this section has been to demonstrate the various types of information available from EQCM studies of monolayer systems, especially monolayers of redox molecules. Given that the application of the EQCM to monolayer assemblies is still in its infancy, and in light of the great current interest in such systems (see Ref. 50 and references therein), it seems clear that EQCM methods will make many important contributions to their understanding.

IV. MULTILAYER DEPOSITION AND DISSOLUTION
A. Introduction

Studies of multilayer depositions and dissolutions have a long history in electrochemistry. Early interest focused on electro-deposition of metals. More recently, the deposition of semicon-ductor materials, the growth and cycling behavior of oxide layers, and the deposition and dissolution mechanisms of electro-chromic films have been popular topics of investigation. While a great variety of phenomena is observed in the electrochemical responses of these diverse systems, many of these studies have probed processes common to all of them, such as mechanisms of nucleation, growth, and dissolution, current efficiency for depo-sition and dissolution, the physical state of the deposit (i.e., roughness, porosity, etc.), and the stoichiometry of the deposit. In this section the use of the EQCM to provide this type of in-formation for selected examples is described.

B. Deposition and Dissolution of Electrochromic
Films of Diheptylviologen Bromide

The highly colored deposits formed by reduction of various salts
of N,N'-disubstituted-bipyridines have been widely investigated
due to their potential applications in electrochromic devices.
Recently, Ostrom and Buttry [45] used the EQCM to study the
mechanisms of deposition and dissolution for films of 1,1'-diheptyl-
4,4'-bipyridinium bromide (diheptylviologen bromide (DHVBr))
under both potential step and sweep conditions. As discussed
in Sec. II, under potential sweep conditions, the observed frequency changes during dissolution were consistent with a pitting
mechanism in which deposit roughness increased during dissolution followed by film disintegration late in the dissolution process.
This conclusion was based on the observation of large apparent
mass loading due to the trapping of supporting electrolyte within
small pores in the film. As will now be discussed, deposition
under potentiostatic conditions also causes large apparent mass
loadings which can be attributed to nucleation processes.

When deposition under potentiostatic conditions occurs, the
rate of growth of the deposit is controlled by the delivery of
reactant to the surface and the nucleation rate, which is usually
a function of the applied potential. Both the charge (Q) consumed in the deposition and the value of Δf provide information
about the deposition. Q reveals the cumulative amount of deposited material (assuming 100% current efficiency), while Δf
gives a measure of the total mass loading from the deposit and
from solvent or supporting electrolyte which might be associated
with the deposit (i.e., trapped in pores or incorporated into the
bulk of the deposit). Thus, comparison of Q and Δf can allow
for evaluation of current efficiency, film composition, and film
roughness under the proper conditions.

For a diffusion controlled deposition under conditions of

100% current efficiency and a uniform, smooth deposit, the equation which describes the time dependence of Δf is:

$$\Delta f = (2 \times 10^6) C_f \; MW \; D^{1/2} C \; t^{1/2} \pi^{-1/2} \tag{21}$$

where D is the diffusion coefficient for the diffusing species (in cm^2/s), C is the concentration (in mole cm^{-3}), t is time (in s), and the other quantities have been previously defined. This equation comes from the combination of Eq. (19) above and the integrated Cottrell equation. Note that MW is the <u>apparent</u> molar mass of the depositing species. It might include contributions from solution which is trapped or incorporated within the deposit or from deposit roughness. When linear plots of Δf versus $t^{1/2}$ and Q versus $t^{1/2}$ are obtained, it is possible to directly compare their slopes to get MW:

$$\frac{S_f}{S_q} \frac{10^6 F}{C_f} = MW \tag{22}$$

in which S_f is the slope of the Δf versus $t^{1/2}$ plot (in Hz $s^{-1/2}$), S_q is the slope of the Q versus $t^{1/2}$ (in C cm^{-2} $s^{-1/2}$). Alternatively, plots of Δf versus Q provide MW directly, as described above. These methods have the advantage of not requiring knowledge of D or C for the determination of MW.

Figure 13 shows the results of a potential step experiment over the first reduction wave of DHV. Curve A is the plot of Δf versus t and curve B is the plot of Δf versus $t^{1/2}$, the microgravimetric equivalent of an Anson plot. Curve B shows that Δf varies with t with essentially the same functional dependence as does Q, in accordance with Eq. (21). Table 1 shows the values for MW obtained for several potential step experiments of this type. Recalling that the peak potential is at ca. -0.55, the data in the table show that the value of MW depends very strongly

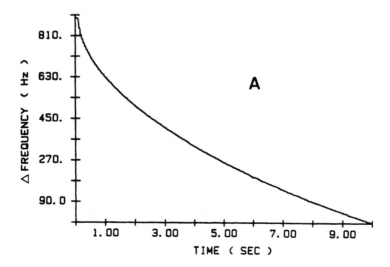

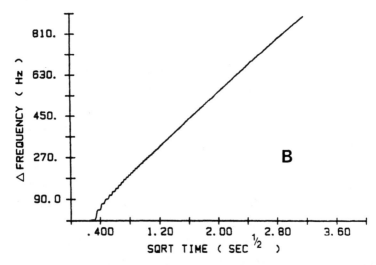

FIG. 13. (A) Relative frequency change versus time for a 10-second step from +0.2 to −0.75 (i.e., over the first reduction wave) for a solution containing 5 mM DHVBr$_2$, 0.3 M KBr, and 5 mM NaOH. (B) Plot of relative frequency change versus the square root of time for the data in (A). (From Ref. 45, with permission.)

TABLE 1

Apparent MWs of the Deposit Formed for
Various Applied Potentials

Concentration of DHVBr$_2$ (mM)	Concentration of KBr (M)	Final potential (V)	MW (g/mol)
1	0.3	−0.580	866
1	0.3	−0.600	731
1	0.3	−0.750	449
5	0.3	−0.535	1205
5	0.3	−0.536	1145
5	0.3	−0.537	957
5	0.3	−0.538	926
5	0.3	−0.750	460

Source: From Ref. 45, with permission.

on the final potential for the step.   Steps far over the wave
give values for MW which are quite close to the molar mass of
DHVBr (434.5 g/mol), while steps to the foot of the reduction
wave give values as much as a factor of three larger than this.
This behavior has been attributed to the number of nucleation
sites for film growth formed during the reduction [45].   For
steps to very negative potentials well past the wave, the number
of nuclei formed is large, so that film growth is relatively uniform.   For steps to the foot of the wave, only a small number
of nuclei are formed.   These nuclei grow hemispherically with
time, and, as a consequence of this, the effective surface roughness of the electrode is increased.   As discussed above, this
increase in surface roughness should result in large apparent
mass loadings and values of MW which are larger than the molar
mass of the depositing species.   This coupling of adjacent solution to the deposited nuclei is schematically depicted in Fig. 14.

Solution "trapped"

between nuclei

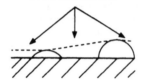

FIG. 14.   Schematic depiction of solution "trapping" between nuclei during the nucleation stage of a deposition.

(Note that the degree to which the solution is coupled to the surface is unknown and should not be inferred from the figure, which serves only to illustrate the effect.)   This model qualitatively predicts that the smaller the overpotential, the smaller the number of nuclei, and, therefore, the greater the extent of coupling of the adjacent solution to the surface.   This will lead to larger values of MW, in agreement with observations.   Quantitative predictions are not possible at the present time, but the development of the theory for such a case would undoubtedly prove useful in the study of nucleation and growth processes for other systems.

The value of MW obtained for steps well past the wave (ca. 460 g/mol) is quite near to the molar mass of DHVBr.   This is strong evidence that at large overpotentials the film deposits relatively uniformly and with little incorporated solvent or supporting electrolyte.   Such information on the solvent content and surface morphology of electrodeposited films is extremely difficult, if not impossible, to obtain using other techniques.   However, these investigations have shown that the EQCM is uniquely well suited for such determinations.

C. Other Systems

Grzegorzewski and Heusler [53] recently published a study in which they used the EQCM to study the mass changes which occur during the redox cycling of MnO_2 films. A particularly novel aspect of this investigation was their construction of a rotating EQCM, thus allowing for the facile control of mass transport to the electrode surface. Through detailed correlations of mass and charge they were able to show that redox cycling of the oxide causes transfer of manganese ions and oxygen ions across the oxide/electrolyte interface in independent processes. They also showed that, in steady state, only electronic currents flow through the oxide, while in transient experiments ionic transport occurs with corresponding changes in the oxide stoichiometry. Using essentially the same analysis as that implied by Eq. (19), they were able to calculate the molar mass of the species associated with oxygen transfer across the oxide/electrolyte interface. A value of 144 g/mol was obtained, suggesting that the oxygen ions were transferred with several molecules of water. As will be discussed below in the context of mass transport during redox events in thin polymer films, the observation of solvent transfer is not uncommon in thin films. However, it is important to note that such transfers should not be considered to be measures of hydration numbers. Rather, these transfers are a consequence of varying thermodynamic activities of water and other components within the film during the redox events [54].

Mori et al. [55] used the EQCM to study the deposition and dissolution processes which occur during the electrochemistry of $HTeO_2^+$ at gold electrodes. The EQCM mass changes, when compared to the electrochemical charges, allowed identification of the direct 4 e$^-$ reduction of $HTeO_2^+$ to Te with the simultaneous deposition of the Te. Using the analysis suggested by Eq. (19), the n value for the reduction was found to be 3.6. This value

was consistently lower than the value of 4 expected, which was
attributed to a competing 2 e⁻ pathway [55]. Further reduction
of Te to H_2Te caused loss of mass due to the dissolution of the
H_2Te. Both this study and the previous one on MnO_2 amply
demonstrate the dramatic enhancement of interpretive capability
towards mechanistic assignments in deposition, dissolution, and
redox switching of thin films provided by the availability of simultaneous, quantitative mass measurements.

V.   POLYMER FILMS

A.   Introduction

Much of the work with the EQCM to date has been on the study
of mass transport processes which accompany redox processes
in polymer films on electrodes. Such systems have been reviewed
[56], and only those aspects of their behavior pertinent to their
study using the EQCM will be discussed here.

Much of the work on polymer modified electrodes (PMEs) has
focused on the process of charge propagation through the polymer film. In the context of these studies, many suggestions
have been made regarding the nature of the rate-limiting processes for the charge propagation [56]. Aside from the physical
diffusion of the redox species through the film and the electron
exchange reaction between redox groups, the transport of solvent and/or supporting electrolyte through the films have been
suggested as potentially rate-limiting processes. Thus, a significant incentive for the study of ion and solvent transport
through polymer films is to understand their role (if any) in the
kinetics of charge propagation.

Ion and solvent transport may also play a role in determining the thermodynamic behavior of PMEs in the sense that the
conditions which produce nonunity activity coefficients for immobilized or incorporated redox groups may be somehow linked
to these transport processes. Thus, the deviations from

Nernstian behavior frequently observed for such systems [56]
might be better understood given detailed information about compositional changes which occur during the redox reactions of the
films.

In the next section, the mass transport processes which
occur during redox reactions in polymer films containing redox
groups are examined. Both organic and inorganic polymers are
discussed. In the following section, the mass transport processes
which occur during oxidation and reduction of conducting polymers are described. The emphasis in these sections is on a demonstration of the types of phenomena which can be observed for
such systems using the EQCM. Examples of such phenomena include current efficiency for deposition in electropolymerization
reactions, extent of swelling of the film and changes in swelling
which occur during redox reactions, ion transport, and changes
in the viscoelastic properties of the deposit. Because changes
in Δf can be measured relatively rapidly (ca. every 10 millisec-
onds) the kinetics of these processes will be accessible in some
instances, although little has been done in this area to date.

B.   Redox Polymers

1.   Poly(vinylferrocene)

The first report of use of the EQCM for measurement of mass
transport processes in redox polymers films concerned the direct
observation of ion and solvent transport in thin films of poly-
(vinylferrocene) (PVF) [57]. Figure 15 [57] shows an example
of the results of such experiments. Curve A shows the CV obtained on a PVF film in 0.1 M KPF_6. Curve B shows Δf versus
E for the scan. The EQCM frequency is seen to decrease substantially during the oxidation process, with reduction causing
the frequency to reattain its original value. Comparison of the
total Δf with the total charge consumed during the oxidation reveals that exactly one PF_6^- anion is inserted into the film for

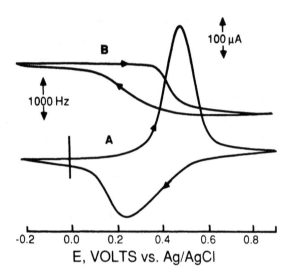

FIG. 15. (A) CV of PVF in 0.1 M KPF$_6$. Scan rate—10 mV/sec. (B) EQCM frequency response for the scan in (A). (From Ref. 57, with permission.)

each electron removed. No evidence for solvent transport is observed, because the EQCM frequency change matches exactly that predicted for the insertion of the appropriate number of anions, based on the charge. Thus, charge compensation is achieved by the influx of anions which serve to neutralize the ferrocenium sites created during oxidation. This behavior is the simplest type that can be observed for such systems, i.e., unidirectional transport with no accompanying solvent. As discussed below, such behavior is more the exception than the rule.

Ward [32] described a study of the mass changes which occur during open circuit reactions of PVF with oxidants in solution. In this study, the oxidation of PVF with I_3^- solutions (which also contain appreciable amounts of I_2) was monitored by the mass change which accompanied the insertion of the charge compensating anion I_3^- into the resulting polycationic film. This method was shown to be very convenient for the determination of the kinetics of such a reaction between a thin film and a solution phase redox reagent. In fact, very high quality data were obtained for the appropriate semilog plots of frequency change versus time, from which a fairly detailed rate law was deduced. This method was also used to monitor the reduction of the nitrate form of PVF^+ by $Ru(NH_3)_6^{2+}$ or $Fe(CN)_6^{4-}$, again following the mass changes due to ion migration driven by the dictates of electroneutrality. In this case, the loss of NO_3^- during reduction of the film and the consequent frequency increases were the experimental observables. It is interesting that for the case of NO_3^- he found that the values for Δf observed were much larger than expected based on the electrochemically harvested charges for these same films. This was attributed to simultaneous solvent and anion transport. This represents a second class of behavior for PVF films, in which the ion transport appears to be accompanied by simultaneous transport of solution.

Figure 16 shows another example of this second type of behavior for PVF [58]. The experiment is similar to that in Fig. 15, except that the solution contains 0.1 M KIO_3. The inner (smaller) curve is the CV and the outer curve is the derivative representation of Δf calculated assuming unidirectional transport of IO_3^- to maintain neutrality and no accompanying solvent transport (see Eq. (20)). When viewed in this way, the extent to which the $d(\Delta f)/dE$ curve is larger than the cyclic voltammetric current is an indication of mass changes which are in excess of those expected for simple anion (IO_3^-, in this case) insertion. Thus, it is clear that the EQCM frequency changes for this system are far in excess of those expected based on the amount of charge consumed during the oxidation, probably as a result of solvent transport in the same direction as the anion transport. However, such results are not unambiguous, and additional information is required to arrive at a definitive description of the

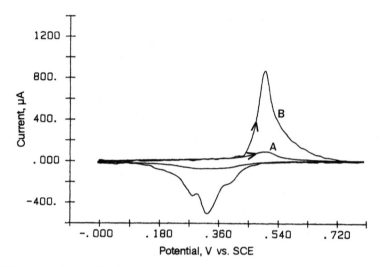

FIG. 16.   (A) CV of PVF in 0.1 M KIO$_3$.   Scan rate—50 mV/sec.
(B) d(Δf)/dE versus E (from Eq. (20)) for the scan in (A).

mass transport processes.   Furthermore, impedance measurements made on this system in both the oxidized and reduced
forms reveal a significant decrease in Q (by a factor of ca. 1.5)
after oxidation in KIO$_3$ electrolyte, indicative of the onset of
viscoelastic behavior for the deposit [58].   Thus, quantitative
calculation of mass changes from Δf is suspect in this particular
instance.   Even with such uncertainty in the quantitative description of the transport processes, it is certainly reasonable
to conclude from such data that anion and solvent transport
occur simultaneously and in the same direction.   This is especially true since the unambiguous observation of the onset of
viscoelastic behavior following oxidation is almost certainly a
consequence of solvent swelling [43].

A third class of behavior is exemplified by the data in Fig.
17 [57].   In this case, the oxidation of the film is carried out
in a solution of 1.0 M NaCl.   Very large increases in frequency

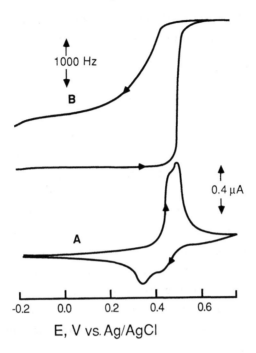

FIG. 17. (A) CV of PVF in 1.0 M NaCl. Scan rate—50 mV/sec.
(B) EQCM frequency response for the scan in (A). (From Ref.
57, with permission.)

(mass loss) are seen to accompany oxidation. These were attributed to the dissolution or delamination of the PVF film following
its oxidation. It has since been discovered that very high molecular weight or cross-linked PVF does not leave the surface
after oxidation. Rather, it behaves similarly to films oxidized
in IO_3^- containing electrolyte, in which extensive solvent swelling occurs [58].

These experiments show that the behavior of these films
follows simple trends which derive from the solubility properties
of the monomeric ferrocenium cation in aqueous solution, in that
more swelling seems to occur in supporting electrolytes in which

the cation would be more soluble. The ability to directly observe such qualitative trends represents a significant step forward towards the goal of rational manipulation of the behavior of such deposits. These data also point out the importance of considering the possibility of simultaneous transport of different species (both solvent and ionic) during the redox reaction. As discussed earlier, there is no a priori reason why these transport processes should occur in any given direction. Rather, the directions in which the various species will flow is entirely dependent on the changes in their thermodynamic activities which occur within the film as a result of the redox reaction, as recently pointed out by Bruckenstein and Hillman [54]. Thus, the simple models of ion insertion and expulsion which derive from considerations of electroneutrality serve only as starting points for a more detailed analysis. Finally, these results reveal a need to more fully understand the factors which dictate the directionality and extent of the various transport processes which can occur in such systems.

2. Nickel Ferrocyanide Films

Thin films of nickel ferrocyanide have been studied for some time in regard to the remarkable sensitivity of the formal potential for the immobilized Fe(III)/Fe(II) redox couple to the identity of the cation of the supporting electrolyte [59]. A recently published EQCM study of such films [60] has provided the first quantitative, unambiguous measurement of solvent transport in thin films on electrodes. Quantitative determination of the extent of solvent transport during the redox process for such films was achieved by measuring the difference in the total mass change (comprised of contributions from both ion and solvent transport) which results from use of isotopically substituted solvent.

As reported in an EQCM study by Feldman and Melroy [61], oxidation of the Fe(II) sites in Prussian Blue films (the iron analog of the nickel ferrocyanide system) results in expulsion of cations from the lattice (and vice versa) as judged by the observation of mass loss during oxidation in EQCM experiments on this system. Discrepancies between the mass loss predicted based on charge consumption and that observed were suggestive of solvent incorporation following cation expulsion. The good stability of the Prussian Blue films [61] and their analogs [59] and their demonstrated permselectivity towards cation transport [59] suggested the possibility of indirect determination of solvent transport using isotopically substituted solvents, since both film stability and permselectivity are essential to this particular experiment.

Figure 18 [60] shows the results of such an experiment. Curves a and b in the upper plate are the cyclic voltammograms for the nickel ferrocyanide film in 0.1 M $CsCl/H_2O$ solution and 0.1 M $CsCl/D_2O$ solution, respectively. The CVs were obtained sequentially by transfer from the H_2O solution to the D_2O solution. The CVs track each other exactly, indicating that equal numbers of Fe sites are electrochemically accessible in these two media. Curves a and b in the lower plate show the EQCM frequency changes observed during the redox reactions in the H_2O and D_2O solutions, respectively. That the frequency increases during the oxidation and decreases during the reduction is expected based on the cation transport model discussed above. However, the mass loss (frequency increase) predicted from the charge consumed in the CVs was considerably greater than that observed, suggesting the presence of some other mass transport process which occurs in the opposite direction. The different values of Δf observed in H_2O and D_2O confirm this conclusion.

Figure 19 shows a schematic representation of the frequency changes observed during the CVs shown in Fig. 18. The electrochemical charge is used to predict the mass loss for the expulsion of Cs^+ ions during the oxidation process, assuming

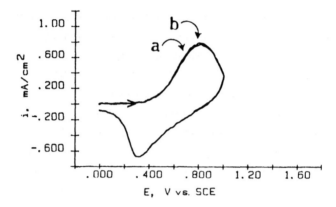

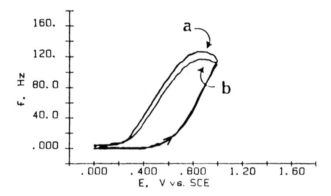

FIG. 18.   Top: (a) CV of a nickel ferrocyanide film in 0.1 M
CsCl/H_2O.   Scan rate—100 mV/sec.   (b) Same as (a) except that
the solution contains D_2O rather than H_2O.   Bottom: (a) EQCM
frequency response for the scan above.   (b)   Same for (b) above.
(From Ref. 60, with permission.)

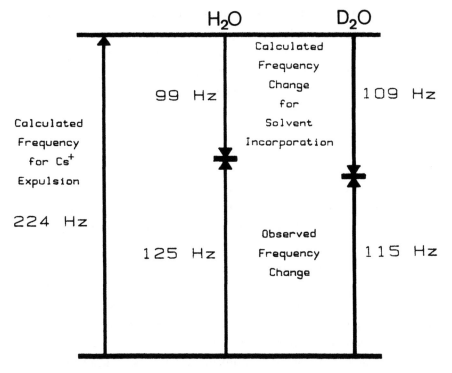

H_2O   **D_2O**

Calculated
Frequency
Change
for
Solvent
Incorporation

99 Hz   109 Hz

Calculated
Frequency
for Cs^+
Expulsion

224 Hz

125 Hz   115 Hz

Observed
Frequency
Change

Fig. 19. Schematic representation of the frequency changes measured and calculated for the CVs in Fig. 18. (From Ref. 60, with permission.)

unidirectional transport (i.e., a permselective film with a transport number for Cs^+ of 1). The predicted value is 224 Hz. The observed values in H_2O and D_2O are 125 and 115 Hz, respectively. The discrepancies between the observed values and the predicted values are 99 and 109 Hz, respectively. If the entire discrepancy were due to solvent transport, then the discrepancy for the D_2O case should be 10% larger than that for the H_2O case, exactly as is observed. Thus, the expulsion of Cs^+ cations is accompanied by the incorporation of significant amounts of solvent. The observed frequency changes are quantitatively predictable based on a model involving cation and solvent transport, verifying the permselectivity of these films (i.e., the absence of anion transport). It is especially significant that such detailed, quantitative determinations of solvent transport can be made in the presence of simultaneous ion transport.

C. Conducting Polymers

1. Introduction

A great deal of study has been done on the redox reactions responsible for the insulator-to-conductor transitions of conducting polymers. In particular, the rate of the switching process has, in some instances, been correlated with the identity of the ionic species in the supporting electrolyte [62,63]. Thus, the identification of the species undergoing transport during the switching reaction is of considerable importance to the ultimate goal of manipulating the properties of these interesting materials for the many applications in which their use is envisioned. The EQCM is an excellent tool for such studies, and has been used in a number of cases already. Additional information available from EQCM measurements is the current efficiency for the electropolymerization process, which is sometimes used to produce thin films of conducting polymers. This is especially useful since, up until recently, the charge consumed in the electropolymerization was used for the indirect determination of the amount of deposited material. This procedure will only be appropriate if the current efficiency is constant when different conditions are used (e.g., supporting electrolyte, solvent, or applied potential or current), a situation which will not be true in general. Also, as will be described below, certain mechanistic questions regarding the electropolymerization can be more easily addressed given direct information on the amount of deposited material produced under different conditions. Finally, the EQCM can provide a qualitative, and sometimes quantitative, picture of the extent of solvent swelling or deswelling which occurs during the switching process for these films. Such data can be very useful in determining the energy density of conducting polymers in relation to their use in charge storage applications.

2. Poly(pyrrole)

The first reported study of ion transport during the switching reaction of a conducting polymer was on poly(pyrrole) [64]. Oxidation of the insulating (neutral) poly(pyrrole) film to its conducting (cationic) state was shown to be accompanied by anion insertion, and reduction by anion expulsion. An exception was found for the case of tetrahydrofuran solvent with lithium perchlorate supporting electrolyte. In this instance, the first oxidation following electropolymerization was accompanied by anion insertion, but subsequent switching was accompanied by Li^+ transport with the ClO_4^- remaining trapped within the film. The lack of ClO_4^- transport was attributed to strong interactions of this ion with the charged sites on the poly(pyrrole) backbone when in its conductive (oxidized) state. This demonstrated the importance of the solvent in mediating the electrostatic interactions of the counterions with the charged sites on the polymer chains, because the effect was only observed for $LiClO_4$ in THF, not in acetonitrile.

A detailed method for determining current efficiencies for electropolymerizations was described by Baker and Reynolds [29], who applied this to a study of the deposition of poly-(pyrrole). Measurement of changes in current efficiency as a function of the reactant concentration was shown to give information valuable in the interpretation of the mechanism of the electropolymerization. They gave a modified version of Faraday's law:

$$Q = \frac{n\,F\,m}{MW} \tag{23}$$

where Q is the charge consumed in the electropolymerization
process, n represents the number of moles of electrons required
to deposit one mole of monomer units, m is the cumulative mass
change for the deposition, and MW is the apparent molar mass
of the deposited species. The value of n can be viewed as a
measure of the current efficiency for the deposition. It is important to note that the value of MW contains the mass of deposited pyrrole units, the mass of any dopant anions which are
contained in the film by virtue of its partial oxidation to the
conductive state at the electropolymerization potential, and the
mass of any solvent contained in the film. The degree of oxidation can be obtained by elemental analysis of a bulk sample of
the film [29], but the amount of solvent incorporation is more
difficult to obtain. In the absence of such information, a value
of zero was used. Then, the value of n is obtained from Q, m,
and the assumed value of MW. This value of n is an upper limit
due to the possible contribution of solvent to the observed mass
change. Since the value of n is inversely proportional to the
current efficiency, this procedure provides a lower limit on the
current efficiency.

For a typical electropolymerization, plots of Q versus m were
linear, with a slope proportional to n [29]. It was found that n
varied with the solution concentration of pyrrole from 5 electrons
per pyrrole unit at 10 mM to 2.3 electrons per pyrrole unit at
100 mM. Measurements over a wide range of concentration
showed that the electropolymerization rate depends on the square
of the pyrrole concentration, in agreement with a proposed mech-
ansim for electropolymerization which involves a second-order
coupling reaction between two pyrrole radical cations [29]. This
set of experiments represents a particularly elegant example of
the power of the EQCM for mechanistic studies of processes which
induce mass changes through deposition or dissolution processes
at the electrode surface.

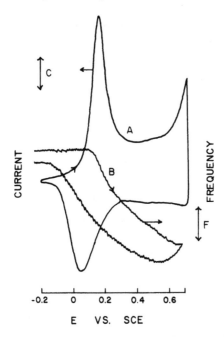

FIG. 20. (A) CV scan during the electropolymerization of PA
in 0.1 M aniline, 1.0 M H_2SO_4. Scan rate—100 mV/sec. C =
100 μA. (B) EQCM frequency response for the scan in (A).
F = 20 Hz. Curve (B) is offset 40 mV to the left with respect
to curve (A). (From Ref. 26, with permission.)

3. Poly(aniline)

The EQCM has been used to study the electropolymerization and
transport processes during switching for thin films of poly(aniline) [26]. Figure 20 shows an example of such an experiment.
Curve A is a CV obtained for a scan recorded in the middle of
an electropolymerization. The sharp anodic wave at ca. 0.15 V
is the characteristic signature of the insulator-to-conductor
transition for this and other conducting polymers. The broad
cathodic wave at ca. 0.05 V is from switching back to the insulating state. These responses arise from the poly(aniline)

which is already present on the electrode surface. The current passed above 0.6 V is from oxidation of anilinium monomer in solution, which ultimately results in the deposition of additional poly(aniline). Curve B shows Δf versus E for the scan in curve A. The decrease in frequency seen to begin at the onset of the anodic switching process and continue until the scan reversal indicates mass gain during the switching and charging processes [26]. As seen by the frequency increase (mass loss) during the return scan, the mass gain from these processes is reversible. Additional mass gain occurs immediately after the scan reversal (between 0.7 and 0.55 V), which is the result of the deposition of additional poly(aniline) as a consequence of the oxidation of anilinium monomer between 0.6 and 0.7 V. The frequency at the end of the scan is lower than its initial value by 10 Hz due to this deposition. This amount of mass gain corresponds to roughly 1.0 nm of new film growth during the scan. The changes in frequency which occur during repetitive scans can also be used to monitor the course of the deposition [26].

Considerable detail regarding the electropolymerization process can be obtained from such measurements. For example, by measuring the total charge consumed during the deposition and comparing this with the total mass change, it is possible to obtain a measure of the current efficiency for the deposition. This was done for deposition of poly(aniline) at constant potential and by potential cycling, and current efficiencies of 15% and 40%, respectively, were calculated assuming a one-electron oxidation of the anilinium monomer for the electropolymerization process [26]. Such measurements provide a relatively simple way to optimize the deposition conditions for maximum current efficiency.

4. Copolymers and Composite Polymer Films

As mentioned earlier, a goal of the study of ion and solvent transport during the switching reaction for conducting polymers (and, in fact, for all thin films which undergo useful redox reactions) is that the information provided from such studies might be used for the rational manipulation of their properties to enhance the switching rates, improve the energy density, or in other ways induce the materials to have more desirable characteristics. Two recent studies serve as examples of the use of the EQCM to aid in the correlation of ion transport with overall charge transport rates. These studies show that, given such information, the rational manipulation of charge transport rates is, indeed, possible.

In the first example, Reynolds and co-workers [63] electrochemically synthesized copolymers of pyrrole and 3-(pyrrol-1-yl)-propanesulfonate in which the pendent sulfonate groups are neutralized with Li^+, K^+, or possibly H^+. Use of the EQCM to monitor ion transport during the switching process for such films demonstrated conclusively that cation transport was the dominant ion transport process during the switching reaction. Thus, the generation of anionic sites for the compensation of the cationic charges along the polymeric chains created by the anodic doping process is the result of expulsion of the cations which had originally been associated with the pendent sulfonate groups. Having definitively demonstrated a system in which unidirectional ion transport occurs, the dependence of overall charge transport rates on the identity of the cationic component of the supporting electrolyte was investigated. Faster charge transport was observed for Li^+-containing electrolytes than for tetrabutylammonium (TBA^+)-containing electrolytes. That the faster rates occur for the cation with the smaller solvated radius (Li^+) is consistent with a model in which the cation mobility is somehow involved in determining the overall charge transport rate. In this particular system, the sulfonated copolymer exhibited slower overall charge transport rates than did poly(pyrrole). However, the study did demonstrate the feasibility of controlling the identity of the ionic species undergoing migration during the switching reaction.

A second example of the application of the EQCM to help
elucidate the directionality of ion transport and the use of such
information to manipulate charge transport rates was a study of
composite films of Nafion and poly(aniline) [65]. Earlier work
on ion transport during switching in poly(aniline) revealed that
anion transport was dominant at pH 1 > 0 [26]. It was thought
that higher charge transport rates could be attained by manipulating the ion transport in such a way as to achieve electroneutrality via proton transport. The reasoning behind this notion
was that the higher mobility of the proton compared to other
ions would allow for faster switching. Towards this end, poly-
(aniline) was electropolymerized within precast films of Nafion,
a perfluorosulfonated polymer with good permselectivity for cation exchange. To the extent that this permselectivity precludes
the rapid transport of anions within the Nafion film, the switching reaction should be accompanied solely by cation transport.
As for the system investigated by Reynolds and co-workers and
described above, these cations would be in the film as the counter
ions of the fixed sulfonate sites of the Nafion matrix. In acidic
solutions, protons would be the only available cations, so they
should be the only species undergoing transport. This situation
is shown schematically in Fig. 21.

Such a model of ion transport based on the maintenance of
electroneutrality is obviously a highly simplified view of the possible mass transport processes which can occur in such films
during redox switching. This is especially true in light of the
recent work by Bruckenstein and Hillman [54], who presented a
thermodynamic analysis demonstrating that electroneutrality alone
cannot account for all of the transport processes in thin films.
They show that, at equilibrium, the net ion and mass transport
processes which result from switching are dictated by the activities of the various species. The implication of this work is
that, for the general case, net transport can occur in either

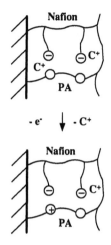

FIG. 21.   Schematic depiction of cation transport during switching in a PA/Nafion composite film.

direction and that the relative molar proportions of the transporting species need not be integer numbers with respect to the number of moles of electrons transferred into or out of the film. However, under transient conditions it is possible that kinetic limitations can be created by the need to achieve electroneutrality [66] (e.g., if the only ions which can migrate to achieve electroneutrality are involved in specific interactions with the fixed sites so that their mobility is decreased), so that manipulation of rates should be possible for some systems by controlling the identities of those species which can fulfill this need.

Contrary to the previous results on simple (noncomposite) poly(aniline) films [26], and in agreement with expectations for the composite films, the switching process in these composite films showed no detectable evidence of anion transport. Figure 22 shows a CV/EQCM experiment for a Nafion film in which poly-(aniline) has been grown to a thickness of roughly half of the Nafion film thickness. In this case, all of the poly(aniline) is

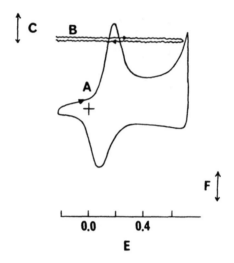

FIG. 22. (A) CV scan during electropolymerization of PA in a Nafion film in 0.1 M aniline, 1.0 M H_2SO_4. Scan rate—200 mV/sec. C = 20 μA. (B) EQCM frequency response for the scan in (A). F = 10 Hz. (From Ref. 65, with permission.)

contained within the interior of the Nafion matrix. As can be seen, scanning through the potential region of the insulator to conductor transition causes no detectable frequency change. This fact was attributed to the predominance of proton transport. Thus, to the extent that electroneutrality in this permselective system is achieved by proton transport with little accompanying solvent transport (an unproven, but not unreasonable, supposition), the observed mass changes should be extremely small due to the small mass of the proton. Another possible explanation derives from the possibility of fortuitous compensation of mass changes which occur in opposite directions. For example, if the Nafion film were not fully permselective, then a net mass change of zero could arise from simultaneous anion insertion and cation expulsion. Such an explanation is not without merit, but the favored model is that of proton transport, since the permselectivity of Nafion films is well known and widely exploited.

An additional explanation for the lack of detectable mass changes during switching in this system would be the loss of mass sensitivity due to shear wave attenuation within the Nafion matrix [65]. If the shear wave were significantly damped within the region of the film close to the electrode surface, mass sensitivity would be lost in regions of the film well past the propagation length of the shear wave. This would occur because only that region of the film which experiences acceleration parallel to the electrode surface can contribute to the EQCM frequency response. This influence of penetration depth of the shear wave into the film has been previously observed in a QCM study of curing in photoresist films [67]. In order to exclude this possibility in the present system, the mass changes in composites in which the poly(aniline) had been grown to thicknesses well beyond the Nafion/solution interface were examined. For such films, that fraction of the poly(aniline) which extends into the solution, past the boundary of the Nafion film, exhibits "normal" mass changes which are characteristic of anion insertion/expulsion in the previously studied (noncomposite) poly(aniline) films. This observation rules out the possibility of significant shear wave attenuation within the Nafion matrix, because in such a situation there could be no EQCM frequency response from mass changes in regions of the film much past the attenuation length. Thus, proton transport still seems the most likely explanation for the observed effects.

In order to more definitively demonstrate the presence of cation transport in these composite films, the mass changes in Cs^+-containing supporting electrolyte solutions were measured. As shown in Fig. 23, these data provide additional evidence for cation transport. (Cs^+-containing electrolyte was used in this experiment to generate the maximum possible mass change from the cation transport process.) Mass loss was observed during the oxidation of the poly(aniline)/Nafion composite (and vice

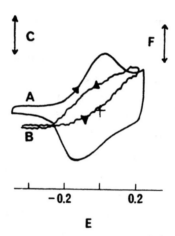

FIG. 23. (A) CV scan of PA/Nafion composite film in 0.1 M
CsCl. Scan rate—200 mV/sec. C = 20 µA. (B) EQCM frequency
response for scan in (A). F = 5 Hz. (From Ref. 65, with permission.)

versa) consistent with Cs^+ expulsion. The mass changes observed for the switching process in this medium were not as
large as expected based solely on cation expulsion to maintain
electroneutrality. Instead, smaller changes were observed, indicative of mixed cation and anion transport and/or cesium and
proton transport (the experiments were conducted at pH3 so the
poly(aniline) would not be too resistive [26,65]) and/or concurrent solvent transport. Thus, considerable ambiguity remains
as to the quantitative details of the mass transport process(es)
for this system. Nevertheless, the data are reasonably convincing of the presence of a considerable component of cation transport for these composite films.

Based on the supposed predominance of proton transport for
these composite films, their charge transport rates were compared to those of the simple (noncomposite) poly(aniline) films
using potential step experiments. The influence of pH was also

examined to test for dependence on the relative proton concentration within the composite film. Depending on the pH, these composite films exhibited charge transport rates as much as a factor of two larger than those of the simple (noncomposite) poly(aniline) films. The enhancement has been attributed to the predominance of proton transport in the composite films [65]. Thus, this study represents an excellent example of how the understanding of charge transport mechanisms can be used in the rational manipulation of these processes.

VI. FUTURE APPLICATIONS OF THE EQCM

This review has attempted to provide an entry into the application of the EQCM to problems of interest to the electrochemical community. As can easily be seen, many types of phenomena can be studied, some of which are not amenable to observation by other methods. In this respect, the EQCM represents a powerful new addition to the electrochemist's repertoire of characterization methods. However, the information obtained from the EQCM measurement is limited in that quantitative information is still only available on mass changes at the surface, and then only when certain criteria are met. It seems likely that as additional theoretical treatments become available, it will be possible to extract some material properties of surface films using the impedance analysis methods described above. In this way, it might be possible to correlate some electrochemical or transport parameters (e.g., permeation, charge transport rates, etc.) with the microscopic relaxation processes upon which these material properties depend.

Despite the relatively nonspecific nature of the information provided by the EQCM, its low cost and simplicity make it quite attractive for the determination of compositional changes at the electrode surface. An additional, significant attribute is the high precision of the measurement of changes in frequency and impedance parameters. The method will prove especially useful when used in conjunction with other methods for interface characterization, such as surface-enhanced Raman spectroscopy, infrared spectroscopy, ellipsometry, etc. In such multitechnique studies, the availability of a tool for such precise mass measurements will undoubtedly be of value, especially when the quantitative aspects of the spectroscopic probes are in question.

ACKNOWLEDGMENTS

We are grateful to the Office of Naval Research for the generous support of our efforts in the application of the QCM to electrochemical problems. Also, many thanks are due to the co-workers whose hard work made this contribution possible.

REFERENCES

1. G. Sauerbrey, Z. Phys. 155:206 (1959).

2. C. Lu and A. W. Czanderna (eds.), Applications of Piezo-electric Quartz Crystal Microbalances, Methods and Phenomena, Vol. 7, Elsevier Science Publishing, New York, 1984.

3. P. L. Konash and G. J. Bastiaans, Anal. Chem. 52:1929 (1980).

4. T. Nomura, Anal. Chim. Acta 124:81 (1981).

5. J. L. Jones and J. P. Meiure, Anal. Chem. 41:484 (1969).

6. J. P. Meiure and J. L. Jones, Talanta 16:149 (1969).

7. T. Nomura, T. Nagamune, K. Izutsu, and T. S. West, Bunseki Kagaku 30:494 (1981).

8. T. Nomura and M. Iijima, Anal. Chim. Acta 131:97 (1981).

9. D. Salt, Hy-Q Handbook of Quartz Crystal Devices, Van-Nostrand Reinhold, Berkshire, England, 1987.

10. R. D. Mindlin, Q. Appl. Math. 19:51 (1961).

11. H. J. McSkimin, P. Andreatch, and R. N. Thurston, J. Appl. Phys. 36:1624 (1965).

12. S. Bruckenstein and M. Shay, Electrochim. Acta 30:1295 (1985).

13. H. K. Pulker and W. Schadler, Nuovo Cimento 57B:19 (1968).

14. H. K. Pulker, E. Benes, D. Hammer, and E. Sollner, Thin Solid Films 32:27 (1976).

15. H. Bahadur and R. Parshad, in Physical Acoustics, Vol. 16 (W. P. Mason and R. N. Thurston, eds.), Academic Press, New York, 1982, pp. 37–171.

16. G. Sauerbrey, Proc. Annu. Freq. Control Symp. 21:63 (1967).

17. G. Sauerbrey, Z. Phys. 178:457 (1964).

18. D. M. Ullevig, J. F. Evans, and M. G. Albrecht, Anal. Chem. 54:2341 (1982).

19. R. M. Mueller and W. White, Rev. Sci. Instr. 39:291 (1968).

20. D. King and G. R. Hoffman, J. Phys. E. 4:993 (1971).

21. D. Salt, Hy-Q Handbook of Quartz Crystal Devices, Van-Nostrand Reinhold, Berkshire, England, 1987, p. 3.

22. R. Schumacher, G. Borges, and K. K. Kanazawa, Surf. Sci. 163:L621 (1985).

23. R. Schumacher, J. G. Gordon, and O. Melroy, J. Electroanal. Chem. 216:127 (1987).

24. O. Melroy, K. K. Kanazawa, J. G. Gordon, and D. A. Buttry, Langmuir 2:697 (1987).

25. M. Benje, M. Eiermann, U. Pittermann, and K. G. Weil, Ber. Bunsen - Ges. Phys. Chem. 90:435 (1986).

26. D. Orata and D. A. Buttry, J. Am. Chem. Soc. 109:3574 (1987).

27. W. Stöckel and R. Schumacher, Ber. Bunsen - Ges. Phys. Chem. 91:345 (1987).

28. M. R. Deakin and O. Melroy, J. Electroanal. Chem. 239:321 (1988).

29. C. K. Baker and J. R. Reynolds, J. Electroanal. Chem. 251:307 (1988).

30. B. Parzen, Design of Crystal and Other Harmonic Oscillators, Wiley-Interscience, New York, 1983.

31. W. Stocker and R. Schumacher, Ber. Bunsen - Ges. Phys. Chem. 91:345 (1987).

32. M. D. Ward, J. Phys. Chem. 92:2049 (1988).

33. S. Gottesfeld, private communication.

34. R. Morrison, Grounding and Schielding Techniques in Instrumentation, Wiley Interscience, New York, 1987.

35. J. G. Miller and D. I. Bolef, J. Appl. Phys. 39:5815 (1968).

36. C. Lu and O. Lewis, J. Appl. Phys. 43:4385 (1972).

37. E. Benes, J. Appl. Phys. 56:608 (1984)

38. A. P. M. Glassford, J. Vac. Sci. Technol. 15:1836 (1978).

39. K. K. Kanazawa and J. G. Gordon, Anal. Chim. Acta 175:99 (1985).

40. K. K. Kanazawa and J. G. Gordon, Anal. Chem. 57:1770 (1985).

41. H. E. Hager, Chem. Eng. Commun. 43:25 (1986).

42. H. Muramatsu, E. Tamiya, and I. Karube, Anal. Chem. 60:2142 (1988).

43. J. D. Ferry, Viscoelastic Properties of Polymers, Wiley Interscience, New York, 1980.

44. C. E. Reed, K. K. Kanazawa, and J. F. Kaufman, preprint.

45. G. S. Ostrom and D. A. Buttry, J. Electroanal. Chem. 256:411 (1988).

46. S. Bruckenstein and M. Shay, J. Electroanal. Chem. 188: 131 (1985).

47. M. R. Deakin, T. T. Li, and O. R. Melroy, J. Electroanal. Chem. 243:343 (1988).

48. S. Bruckenstein and S. Swathirajan, Electrochim. Acta 30:851 (1985).

49. J. J. Donohue, L. J. Nordyke, and D. A. Buttry, Chemically Modified Surfaces in Science and Industry (D. E. Leyden and W. T. Collins, eds.), Gordon and Breach, New York, 1988, p. 377.

50. J. J. Donohue and D. A. Buttry, Langmuir 5:671 (1989).

51. T. Saji, K. Hoshino, and A. Aoyagui, J. Am. Chem. Soc. 107:6865 (1985).

52. J. J. Donohue and D. A. Buttry, manuscript in preparation.

53. A. Grzegorzewski and K. E. Heusler, J. Electroanal. Chem. 228:455 (1987).

54. S. Bruckenstein and A. R. Hillman, J. Phys. Chem. 92: 4837 (1988).

55. E. Mori, C. K. Baker, J. R. Reynolds, and K. Rajeshwar, J. Electroanal. Chem. 252:441 (1988).

56. R. W. Murray, Electroanalytical Chemistry (A. J. Bard, ed.), Marcel Dekker, New York, 1984, p. 191.

57. P. T. Varineau and D. A. Buttry, J. Phys. Chem. 91:1292 (1987).

58. P. T. Varineau and D. A. Buttry, manuscript in preparation.

59. S. Sinha, L. J. Ames, M. H. Schmidt, and A. B. Bocarsly, J. Electroanal. Chem. 210:323 (1986).

60. S. J. Lasky and D. A. Buttry, J. Am. Chem. Soc. 110: 6258 (1988).

61. B. J. Feldman and O. R. Melroy, J. Electroanal. Chem. 234:213 (1987).

62. R. M. Penner and C. R. Martin, J. Phys. Chem. 92:5274 (1988).

63. J. R. Reynolds, N. S. Sundaresan, M. Pomerantz, S. Basak, and C. K. Baker, J. Electroanal. Chem. 250:355 (1988).

64. J. N. Kaufman, K. K. Kanazawa, and G. B. Street, Phys. Rev. Lett. 53:2461 (1984).

65. D. Orata and D. A. Buttry, J. Electroanal. Chem. 257:71 (1988).

66. J. M. Saveant, J. Phys. Chem. 92:4526 (1988).

67. H. E. Hager, G. A. Escobar, and P. D. Verge, J. Appl. Phys. 59:3328 (1986).

OPTICAL SECOND HARMONIC GENERATION AS
AN IN SITU PROBE OF ELECTROCHEMICAL INTERFACES

Geraldine L. Richmond

Department of Chemistry
University of Oregon
Eugene, Oregon

I. INTRODUCTION
A. Background

Since the first observation of the generation of optical harmonics
in 1961 from passage of a pulsed ruby laser beam through crystalline quartz [1], optical second harmonic generation (SHG)
and the encompassing field of nonlinear optics has shown rapid

growth. Not only has this growth been in the fundamental
understanding of the phenomena in bulk materials [2], but major
advances in the application of second harmonic generation to a
variety of important technologies have also been witnessed. The
field continues to grow at a rapid pace as new optical materials
and devices are synthesized and designed. A subset of this
field, surface second harmonic generation, has also progressed
but at a somewhat slower pace until recently, due largely to difficulties in describing and understanding fundamental issues regarding the source of the nonlinear response at such an abrupt
junction. However, in the past few years this subfield also appears to have gained momentum as scientists from a diverse
spectrum of disciplines interested in interfacial phenomena have
begun to investigate the usefulness of SHG to their respective
areas.

One of the primary reasons for interest in this area is the
inherent sensitivity of this nonlinear optical technique for examining surface and interfacial regions. This sensitivity arises
from the selection rules for this second order process which isolate the production of SH light to materials or regions in the
media where inversion symmetry is lacking. The interfacial region between two adjoining centrosymmetric media provides this
asymmetry naturally. The result is the production of SH light
with characteristics which depend upon, and can potentially
provide information about, the electronic and structural properties of that atomically thin interfacial junction. This is of significance for electrochemical measurements since one can specifically probe on a monolayer level the interfacial region without
interfering contributions from the bulk of the electrode or the
electrolyte. A second advantage which has yet to be explored
to any large extent in electrochemical measurements is the ability
to explore time-dependent phenomena. Since pulsed lasers are
generally used as the incident beam, one has the possibility of

measuring events in the range of 10^{-6} to 10^{-12} sec time regime which is generally inaccessible by conventional electrochemical methods.

The focus of this review will be on electrochemical studies in the field of surface SHG. The review begins with a brief overview of the theory behind the phenomenon. Following this discussion will be a description of some common experimental procedures for making the nonlinear optical measurements. The primary portion of the review will focus on experimental electro-chemical studies ranging from the pioneering work in which the SH response from a surface undergoing an oxidation-reduction cycle is examined, to more recent studies of the use of SHG to monitor surface structural properties on single crystal electrode surfaces in solution and in UHV. Many of the studies to be discussed exploit the unique properties of the electrochemical interface to derive fundamental information about the source of the nonlinear optical polarizability from an interfacial region. Others use the technique to provide insight into static and dynamic properties of the electrochemical double layer. Integrated into this discussion are time-resolved measurements of various adsorption and deposition kinetics. The review will conclude with a brief discussion of future directions in this and related fields. For a more extensive review of surface SHG in the above-mentioned areas beyond the electrochemical studies, the reader is referred to two recent summaries [3,4].

B. Historical Perspective

Bloembergen and co-workers were the first to make a theoretical prediction [5] and an experimental observation [6] of SHG in 1962 and 1963, respectively. Other theoretical and experimental works which then followed throughout the decade were mainly focused on understanding the phenomenon at metallic surfaces [7–21]. Efforts continued in the 1970s [22–35] with the most

significant progress made in the theoretical area, as will be described in the next section. Most of the surface work during these
decades was on the investigation of the nonlinear optical phenomenon
itself with little concern toward applicability. Unfortunately,
most of the experimental studies in this area were hampered by
inadequate surface preparation and analysis methods, which led
to ambiguities in the interpretations.

In the early 1980s, a surge in interest in the technique occurred as a consequence of the pioneering work of Shen and co-
wokers on enhancements in surface SHG observed on electrochemically roughened silver surfaces [36]. At this time there
was also a large interest in surface enhanced Raman scattering
(SERS) from roughened silver surfaces [37]. Due to the com-
monality of the surfaces studied and the large enhancements observed for both, interest in the SHG studies rode largely on the
coattails of the SERS work. However, unlike SERS, it soon became apparent that in an electrochemical setting, the SH response
in simple electrolytes was originating primarily from the surface
itself and not necessarily from the surface adsorbed molecular species
whose vibrational spectra the Raman scattering measurements examined. Although information about molecular structure cannot
be obtained when SHG is used in this single wavelength mode,
unlike SERS, surface SHG is applicable to a wide variety of
roughened and unroughened metals, semiconductors, and insula-
tors.

Over the past 10 years, the field of surface SHG has matured.
A variety of metal and semiconductor surfaces have been examined under ultra high vacuum (UHV) conditions [38–51] as
well as in more reactive settings such as in contact with other
metals [52–64], polymers [65], and liquids [13,36,66–103].
Both static and time-resolved measurements have been reported.
In addition, recent work has demonstrated its applicability to
studying liquid/liquid and liquid/air interfaces [104–113]. Many

of these applications have been exploratory in nature, since the fundamental understanding of the source of the nonlinear optical polarizability lags far behind many of the experimental studies. However, these studies have provided, and will continue to contribute, valuable input for developing fundamental theories to describe surface SHG.

II.   THEORETICAL CONSIDERATIONS
A.   General Theory

Since the first theoretical treatment of second harmonic generation by Bloembergen and Pershan in 1962 [5], considerable effort has been expended in this area to describe the phenomenon of second harmonic generation. Although the theory of second harmonic generation in crystals lacking inversion symmetry is fairly well developed, progress in describing SHG from an interfacial region lags far behind. Most of the difficulty in this latter case lies in describing the source of the nonlinear polarizability. Once an expression for this source is determined, the actual fields generated by the source can be found [114]. Bloembergen and Pershan originally solved the problem for a generalized dipole source [5]. Since then, many detailed models have been attempted for calculating the microscopic source of the nonlinear polarization at the surface of various media [8,9,12,19,12, 19,22,23,31,115–117]. The work described here will primarily be directed towards metallic surfaces, since this is where most of the electrochemical work has been done.

A generalized picture of SHG from an interface is given in Fig. 1, in which the surface atomic or molecular layers have optical properties which are distinct from those of the bulk. The interfacial region consists of a multilayer geometry where the surface layer of thickness d and linear dielectric constant ε_2 is sandwiched between two isotropic bulk media with respective linear dielectric constants ε_1 and ε_2. A monochromatic

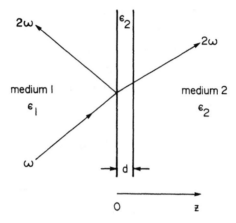

FIG. 1.   Second harmonic generation from an interface between
two isotropic media.   The z direction is taken as zero at the interface between the media and is positive going into medium 2.
(From Ref. 2.)

plane wave at frequency ω incident from medium 1 induces a
nonlinear source polarization in the surface layer and in the
bulk of medium 2.   This source polarization then radiates, with
harmonic waves at twice the incident frequency (2ω) emanating
from the boundary in both the reflected and transmitted directions.   In this model, medium 1 is assumed to be strictly linear.

Because of the absence of inversion symmetry at the interface, an electric dipole contribution to the second-order nonlinearity is created.   Furthermore, the structural discontinuity
at the surface is also responsible for the field discontinuity,
which leads to additional quadrupole type contributions [118,119].
The standard method employed is to treat the nonlinear polarization at the surface (at $z = 0^+$) as a dipole sheet which has a
surface susceptibility tensor, $\chi^{(2)}$.   This is equivalent to considering the three-layer system shown in Fig. 1 and allowing
the surface layer thickness to approach zero.   The expression
for this surface polarization then takes the form [2]:

$$\overline{P}_s(2\omega) = \chi^{(2)}_s \delta(z) : \overline{E}(\omega)\overline{E}(\omega) \qquad (1)$$

where $\delta(z)$ is a δ-function at $z = 0^+$.

For completeness, higher order contributions from the bulk of medium 2 must also be considered. Whereas the second-order nonlinear susceptibility given by $\chi^{(2)}$ vanishes in the bulk of medium 2 because of inversion symmetry, it does not vanish if the higher order magnetic dipole and electric quadrupole source terms are considered. These contributions arise from the non-local response of the medium and are proportional to the first spatial derivative of the field:

$$\overline{P}(2\omega) = \overline{Q}(2\omega) : \overline{E}(\omega)\, \nabla\overline{E}(\omega) \qquad (2)$$

Q is a fourth rank tensor of multipole terms. Symmetry dictates the number of independent nonzero tensor elements. For iso-tropic media there are three nonvanishing tensor elements. The ith component of the induced second-order polarization may then be written (in the notation of Bloembergen et al. [191]):

$$P_i(2\omega) = (\delta - \beta - 2\gamma)[\overline{E}(\omega) \cdot \nabla]E_i(\omega) + \beta E_i(\omega)[\nabla \cdot \overline{E}(\omega)]$$
$$+ \gamma\nabla_i[\overline{E}(\omega) \cdot \overline{E}(\omega)] \qquad (3)$$

The coefficients δ, β, and γ relate to the independent tensor elements, and are frequency-dependent material parameters. For cubic media, another term, $\zeta E_i(\omega)\nabla_i E_i(\omega)$, appears in this ex-pression, as will be discussed later. By application of Maxwell's equation, the above expression can be transformed to:

$$\overline{P}(2\omega) = (\delta - \beta)[\overline{E}(\omega) \cdot \nabla]\overline{E}(\omega) + \beta\overline{E}(\omega)[\nabla \cdot \overline{E}(\omega)]$$
$$+ (2i\omega/c)\gamma[\overline{E}(\omega) \times \overline{B}(\omega)] \qquad (4)$$

where the first two terms can be seen to have electric quadru-pole character, while the last term is magnetic dipolar in nature. With excitation by a single plane wave, the first term vanishes, whereas in a homogeneous medium the second term vanishes by

Gauss's Law. For the third term, the induced polarization is along the propagation direction and thus can only radiate at the discontinuity of the surface.

Combining the surface and bulk higher order contributions leads to the expression for the polarization of an isotropic medium as:

$$\overline{P}_{s,eff}(2\omega) = (\delta - \beta)[\overline{E}(\omega) \cdot \nabla]\overline{E}(\omega) + \beta\overline{E}(\omega)[\nabla \cdot \overline{E}(\omega)]$$
$$+ (2i\omega/c)\gamma[\overline{E}(\omega) \times \overline{B}(\omega)] + \chi^{(2)}_s \delta(z): \overline{E}(\omega)\overline{E}(\omega)$$

$$(5)$$

Once the form of the nonlinear polarization is established, the radiated SH fields are found by solving the wave equation [19]:

$$\nabla \times \nabla \times \overline{E}_2 - (4\omega^2/c^2) \, \epsilon(2\omega)\overline{E}_2 = (16\pi\omega^2/c^2)\overline{P}^{NL}(2\omega) \qquad (6)$$

given the new boundary conditions imposed by the presence of the dipole layer. $\overline{P}^{NL}(2\omega)$ has the form given by $\overline{P}_{eff}(2\omega)$ in Eq. (5).

The expression for the reflected SH signal from excitation of a plane wave of frequency ω and polarization $e(\omega)$ can then be written as [2,120,121]:

$$I(2\omega) = \frac{32\pi^3\omega^2 \sec^2\theta_{2\omega} |\overline{e}(2\omega) \cdot \chi^{(2)}_{s,eff}: \overline{e}(\omega)\overline{e}(\omega)|^2 I^2(\omega)}{c^3 A} \qquad (7)$$

In this equation, $\theta_{2\omega}$ represents the angle of the radiated SH light with respect to the surface normal, $I(\omega)$ is the pump intensity, and A is the cross-sectional area of the beam. The vectors $\overline{e}(\omega)$ and $\overline{e}(2\omega)$ are related to the unit polarization vectors $\hat{e}(\omega)$ and $\hat{e}(2\omega)$ in medium 2 by Fresnel coefficients. The effective surface nonlinear susceptibility $\chi^{(2)}_{s,eff}$ incorporates the surface nonlinear susceptibility $\chi^{(2)}_s$ and the bulk magnetic dipole contributions to the nonlinearity. The result simplifies

since, for isotropic media, there are only three nonzero inde-
pendent elements of $\chi^{(2)}_s$. These are $\chi^{(2)}_{s,zii}$, $\chi^{(2)}_{s,izi}$ =
$\chi^{(2)}_{s,iiz}$, and $\chi^{(2)}_{s,zzz}$ where i = x,y. This final expression
for the reflected second harmonic intensity is general and does
not take into account the microscopic source of the nonlinear
polarizability from a given medium.

The microscopic source of the polarizability has been examined in a large number of theoretical studies. For centrosymmetric materials of insulating or free electron behavior, the
bulk nonlinear response is relatively well understood and can be
calculated from the lowest order nonlocal terms with reasonable
accuracy. Calculations of the surface contribution continue to
be a problem due to a number of factors including inadequate
knowledge of surface electronic structure. Complicating the
situation is the fact that it is not possible to separately determine the elements of the surface susceptibility and the bulk
constants [122,123]. Estimates of the relative magnitude of the
surface and bulk terms have been made by Shen based on surface layer thickness and the size of the unit cell of the material
[2]. The results suggest that the surface contribution is of
equal or greater magnitude than that of the bulk systems examined. In studies to be discussed later, experimental parameters have also been varied in attempts to enhance the surface
components over the bulk and thereby isolate the surface effects.
These parameters include polarization and wavelength selection
and surface roughening.

B. Nonlinear Response from Metallic Surfaces

Although much effort has been expended over the past decade
in deriving a theoretical description of SHG in reflection from a
metal surface, progress has been slow. At the center of the
problem is the derivation of expressions for the two types of
source currents [22]: a bulk current density which extends to

approximately a skin depth into the metal, and the surface current density which extends only a few Fermi wavelengths into the metal. Of the two surface currents which exist, one normal and one tangential to the surface, the former is the most difficult to calculate because of the discontinuity at the interface and the rapidly varying normal component of the electric field. It is also predicted to be the most sensitive to the details of the surface [124]. For these reasons it has been the focus of much debate over the past 20 years. A brief summary of some of the work done in calculating the magnitude of these source currents is given below.

Various models have been used to calculate the nonlinear response, the simplest being the isotropic free-electron gas mode used initially by Jha [8]. The second-order nonlinear response using such a model has been calculated in a number of early works through classical and quantum mechanical methods [9,12, 19,22,23,29,116]. In the early years, of particular significance is the work of Rudnick and Stern [22], who used this model in the long wavelength local limit to show that the longitudinal current can be correctly calculated but that difficulties arise in this limit for the perpendicular current since the normal component of the electric field at the surface varies over a few Fermi wavelengths. In an attempt to describe the magnitude of this normal component, they introduced the dimensionless parameter $\alpha(\omega)$, where α is the integrated normal component of the SH surface polarization. The parameter was generally believed to be approximately unity, suggesting that the perpendicular surface current gives only a minor contribution to the total SH signal.

The hydrodynamic model of the electron gas has been employed to calculate the nonlinear source currents in a number of studies by Sipe et al. [31,125] and Corvi and Schaich [126]. In this model collective electron motions are considered, a deviation from the free-electron model mentioned above. Both groups suggest interesting frequency dependencies in the longitudinal SH response near or below the bulk plasma frequencies.

In a series of works by Keller [127–132], the second order response from free-electron-like metals has been examined by employing extensions of linear theories on the optical response of metals. A semiclassical infinite barrier model is used in which the barrier seen by the electrons at the surface is sharp and infinitely high [127,128]. The results reduce to the hydrodynamic theories of Sipe et al. [31,125] and Corvi and Schaich [126] in the collisionless limit.

Both the free-electron model and the hydrodynamic model are valuable as starting points but unfortunately are too crude to give a quantitative description of the dynamics of the electrons near a surface [22,31,126]. Each suffers from the unrealistic assumption of a constant distribution of electrons up to the surface which drops abruptly to zero. A modification on these models is the quantum mechanical approach of Weber and Leibsch [124,133] in which a density-functional theory has been used to calculate the longitudinal response from a simple metal in the low frequency limit. Electron–electron interactions in this jellium model are described within the local density approximation of exchange and correlation which leads to a smoothness of the electron distribution in the surface region. The results demonstrate that the surface current is located predominantly in the exponential tail of the equilibrium density and that the SH signal should therefore be very sensitive to surface conditions.

Quantitative agreement with the density functional theory of Weber and Leibsch has come from the work of Chizmeshya and Zaremba [134], who have employed a generalized Thomas Fermi model in which the quantum mechanical kinetic energy of a system of independent fermions is represented as a function of the electron density. With a correct description of ground-state properties, this method may be extended to provide a hydro-

dynamic description of time-dependent phenomena with results
similar to those obtained in a fully quantum mechanical description.

In more recent work by Leibsch [135], the frequency dependence of SHG at the metal surface is calculated by use of
the time-dependent density functional approach. The conclusion
drawn from the work is that the overall magnitude of the normal
component, $\alpha(\omega)$, is much larger than unity for all bulk densities examined, placing it in relative magnitude above the bulk
and parallel surface contributions.

Aers and Inglesfield [136] have recently published work
which is a significant departure from the previous jellium models
and could be particularly valuable for understanding potential
induced SHG from metal surfaces in the electrochemical cell. In
this work, the electronic structure of Ag(001) is calculated self-
consistently with electric fields of different strengths. In a low
frequency field, this nonlinear effect contributes to the longitudinal current normal to the surface and is thus important in SHG
studies. The work is also correlated with the electroreflectance
work of Kolb and co-workers [137–139] on this system.

As is evident from the description above, a large gap currently exists between theory and experiment. Much of the experimental data needed to evaluate the success of various models
does not exist, i.e., angular and wavelength-dependent studies
and examinations of a wide variety of materials. Furthermore,
most of the theoretical calculations are based on free-electron-
like models, which are thus unrealistic for most materials either
in vacuum or in the electrochemical cell. Therefore, most of the
experimental work to be discussed below will be described in
terms of the phenomenological macroscopic models which provide
the most insight to date.

III.   EXPERIMENTAL METHODS

The experimental techniques used for making the SH measurements in the electrochemical cell are relatively simple.   The
basic optical components consist of a pulsed or continuous wave
(cw) laser, an electrochemical cell which has an appropriate
optical geometry, and detection optics.   A schematic diagram il-
lustrating the standard optics which can be used when the experiment is operated in a reflection mode is shown in Fig. 2.
Linearly polarized light of frequency ω strikes the surface at
an angle θ from the surface normal.   The second harmonic light
at 2ω is produced at an angle determined by the Fresnel factors
at that wavelength.   This angle is generally near the angle of
specular reflection of the fundamental beam, allowing one to use
the fundamental as an approximate means of alignment of the SH
light to the detector.   The resulting SH beam is analyzed for

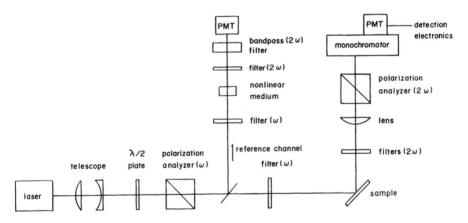

FIG. 2.   Schematic diagram of the second harmonic generation
experimental apparatus with the sample in the reflection geometry.
The polarization analyzers are set to transmit p-polarized light
at the frequency labeled in the figure.   The (ω/2ω) filters transmit the (fundamental/harmonic) light while blocking the (har-
monic/fundamental) light.

polarization and is detected after rejection of the reflected fundamental beam by color filters, spatial separation optics, or a monochromator. The latter is important for verifying the monochromaticity of the measured light, particularly when luminescene is an interfering problem. Depending on the signal levels, either a photon counter or gated electronics can be used for collecting the signal from the phototube. To compensate for fluctuations in the laser incident intensity, a fraction of the beam can be extracted with a beam splitter, sent through a doubling medium, and detected with a reference detector. The normalized SH response is then collected as the potential is swept.

Although either a cw or a pulsed laser source can be used, the high peak powers of the latter is most desirable. Either a single wavelength source or a tunable dye laser coupled to a pump laser can be used depending on the extent of the tunability of frequency desired. Use of a tunable laser source allows the experimentalist to probe optical resonance effects by varying ω or 2ω over an electronic transition of the material or the adsorbing substrate. Q-switched lasers are often the laser of choice for such experiments since they most readily provide sufficient pulse energies for detectable signals. A 10–30 Hz Nd:YAG laser is the most commonly used Q-switched source. As an example of the conversion efficiency expected, for a laser beam intensity of $I(\omega) = 10$ MW cm^{-2} for 1060 nm pulses of 10 nsec duration, a response of 10^3 SH photons/pulse would be predicted from Eq. (7). This calculation is for a material which has a second-order nonlinear susceptibility of approximately 10^{-15} esu [2].

SH experiments using a cw mode-locked Nd:YAG laser with [81] and without [59] Q-switching in conjunction with photon counting detection have also been reported. One advantage that the mode-locked system has over the Q-switched device is that the high repetition rate allows continuous interrogation of a transient event at a surface. The experimental time resolution for the higher repetition rate mode-locked system is about 13 nsec, whereas it typically ranges from 30 to 100 msec for the Q-switched laser. This approach is ideal for studying electro-chemical phenomena following a fast potential step [59,95]. Time-resolved pump-SH probe studies with subnanosecond time resolution have been reported in a number of studies [140–144]. The pump-probe configuration has the potential to provide dynamic information following an optically induced temperature jump or photo-excitation of the material.

Although the reflection geometry is the simplest and most versatile, studies have also been performed using an attenuated total reflection (ATR) geometry. Using the configuration shown in Fig. 3, the ATR geometry allows coupling of surface plasmon waves to the incident electric field resulting in an enhancement of several orders of magnitude in SH response. Two designs are generally used. In the Otto configuration a gap exists be-tween the metal and the coupling prism [145]. In the Kretsch-mann configuration shown in Fig. 3, the gap is filled with an index matching fluid or the metal is deposited directly on the prism [145]. The incident light launches a plasmon surface polariton (PSP) along the surface, defined as the x direction, when the x component of the incident wave vector equals that of the surface polariton. Provided the phase matching condition is met, a second harmonic photon is produced at the plasmon angle of incidence, Θ_p, such that the x component of the SH wave vector equals twice that of the surface polaritons [103]. The efficiency of the coupling depends on the thickness of the metal film or the air gap in the respective configurations. These sample configurations have been most generally used for study-ing thin deposited films.

Several variations on the standard reflection geometry are used for studies to be described which are aimed at measuring the surface symmetry and order of crystalline materials in situ.

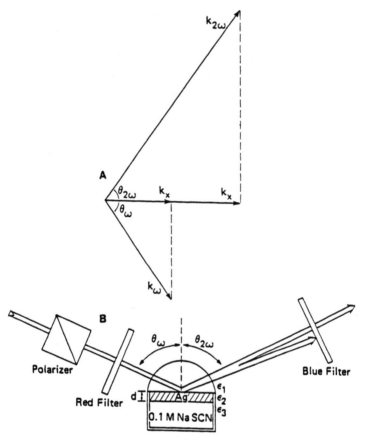

FIG. 3. (A) Wave vector diagram for surface plasmon-enhanced SHG in the Kretchmann configuration. The incident wave vector k_ω excites a PSP with wave vector k_x on the surface. The second harmonic beam is reflected with a wave vector $k_{2\omega}$. (B) Schematic of the experiment. The incoming light is p-polarized. d is the thickness of the silver film, ε_1 to ε_3 are the dielectric constants for the prism, silver, and solution, respectively. (From Ref. 61.)

These experiments involve biasing the surface at a fixed electrode
potential and rotating the crystal surface about the surface normal for fixed incoming fundamental and outgoing harmonic beam
polarizations, as shown in Fig. 4. The result is a modulation
in the SH response as a function of rotation angle, known as
rotational anisotropy, which reflects the symmetry of the crystal
surface. Another approach is to fix the angle of rotation as
well as the input and output polarizations and to sweep the potential throughout the region of interest. Various combinations
of input and output polarizations can be used in order to acquire
the information desired. The rotational anisotropy in the SH
response can also be measured by fixing the electrode position
and having the fundamental light strike the sample at normal
incidence. Instead of rotating the surface, the polarization of
the incoming beam is rotated and the SH light generated is analyzed along two orthogonal directions [146].

The orientation of a molecular monolayer relative to the surface normal can be measured by SHG by analyzing the polarization dependence. Although this has not been used to any large
extent in electrochemical studies due to the overwhelming contribution from the substrate, for conditions in which the adsorbate

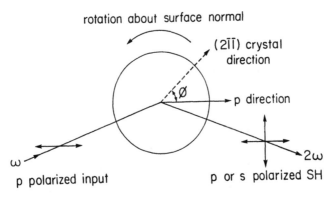

FIG. 4. Schematic diagram of the SH rotational anisotropy experimental geometry for a crystal surface of (111) orientation.

$\chi^{(2)}$ dominates the substrate, these experiments have potential to be very valuable. In the simplest arrangement, the angle of the molecular axes can be experimentally determined from measurements of the ratio of $I(2\omega)$ and $I^2(\omega)$ for various input and output polarizations [147]. In practice, laser fluctuations make it difficult to obtain accurate results. To overcome such difficulties, a simple nulling scheme is used in which the output polarizer is set to block the SH output. Under these conditions, a value for the relative orientation of the molecular axes can be readily obtained.

IV.   EXPERIMENTAL APPLICATION
A.   SHG During an ORC

After the early report of SHG from silver electrode surfaces in 1967 [13], nothing appeared on this topic until 1981 when Shen and co-workers reexamined the Ag/KCl system more extensively [83]. They reported that an electrochemically roughened Ag surface enhanced the SH signal by 10^4 relative to an unroughened silver surface. Figure 5 shows an example of the results obtained using a 1064 nm incident beam with the 532 nm SH response collected during the ORC. Upon the formation of AgCl on the surface (+0.08 V), a strong rise in the SH intensity is observed. With reversal to more anodic potentials (−0.30 V), the silver chloride is reduced back to the silver surface and the SH signal drops. The significance of the work was the demonstration that the electric dipole contribution from the surface layer was at least as strong as the electric quadrupole and magnetic dipole contribution from several hundred monolayers in the bulk. This claim was based on their ability to obtain measurable signals from fewer than 3–4 monolayers of AgCl on the surface formed during the ORC. They also showed a sensitivity to adsorption of pyridine at negative potentials where this adsorption is anticipated to occur (Fig. 5).

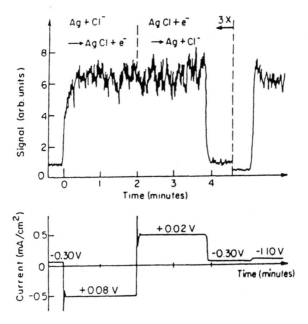

FIG. 5. Current and diffuse second-harmonic reflection as a function of time during and after an electrolytic cycle. The voltages listed on the lower curve are V_{Ag-SCE}. 0.05 M pyridine was added to the 0.1 M KCl solution following the completion of the electrolyte cycle. (From Ref. 83.)

Due to the large interest at the time in SERS from molecules adsorbed on these roughened silver electrode surfaces, several of these early studies were aimed at comparing the variation in the SH and SERS intensities during an ORC as well as developing a description for the possible enhancement mechanisms. For SERS, two distinct contributions are generally acknowledged to explain the observed enhancement in the Raman signals for vibrational modes of the adsorbed molecule [37]. The first type is a chemical interaction between the adsorbed molecules and the substrate which can lead to an intrinsic Raman cross section that is different from that of the isolated molecule. The "local field

enhancement mechanism" involves the enhancement of incoming
and outgoing optical fields resulting from local-plasmon resonances
and corona effects in the rough surface features. This latter
mechanism should be general for any surface optical process,
the presence of the roughened metal producing a local field
strength at the interface which greatly exceeds the applied field.
This latter mechanism appears to be dominant in the SHG from
roughened metal surfaces [84,148].

Shen and co-workers [84] and Murphy et al. [88] both examined the correlation between the SHG and SERS response from
Ag in K_2SO_4, KCl, and CN^- containing solutions. The former
study focused on developing a general formalism used to predict
the observed enhancement in radiated power for the two optical
processes. They concluded that the dominant mechanism responsible for the enhancement in the SH response was a local field
effect. Furthermore, their comparisons of the two optical techniques led them to propose that SHG was more sensitive than
SERS to molecular orientation and symmetry.

Appearing at nearly the same time was a similar study by
D. V. Murphy et al. [88] in which the SH and SERS behavior
for silver in similar electrolytes was compared. They also noted
the role of surface roughness in enhancing SH signals from silver
surfaces undergoing an ORC. Another important observation
made in this study was an irreversible change in the surface
which occurred upon illumination with pulsed lasers. When they
illuminated the surface during cycling to levels of 470 mJ/cm^2
per pulse at 220 pps ($10W/cm^2$), they found noticeable changes
in the scanning electron micrographs relative to an unilluminated
the surface. They proposed that this was due to photon induced
reduction of the AgCl or Ag_2SO_4 on the surface.

The SH response from the Ag/K_2SO_4 system was examined
in more detail by Richmond [72,87]. This study was aimed at
correlating intensity changes in the SH response during the ORC

with different electrochemically formed surface species. It was found that changes in the SH intensity during electrochemical formation of Ag_2SO_4 were highly pH dependent. This pH effect was attributed to the formation of oxides on the surface which consequently control the magnitude of the SH signals upon an ORC. These potential dependent changes in the SH response were previously attributed to changes in orientation and symmetry of the Ag_2SO_4 at the surface [84]. Figure 6 shows an example of this effect for Ag cycled in solutions of K_2SO_4 at various pHs. At pH values of 5 or greater, AgO forms an insulating layer on the surface near +0.3 V and diminishes the

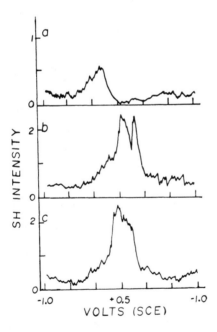

FIG. 6.   SH intensity from 0.1 M sulfate containing electrolytes at selected pH values.   (a) 0.1 M K_2SO_4 adjusted to pH 11.3; (b) 0.1 M K_2SO_4 at pH 5.0; (c) 0.1 M K_2SO_4 at pH 3.2.   (From Ref. 87.)

SH response from the Ag_2SO_4 (Fig. 6a and b). However, as
the pH is lowered and the oxide becomes more soluble, the strong
signal from the Ag_2SO_4 returns (Fig. 6c). It was suggested
that a similar oxide could be forming on the surface when the
Ag is cycled in basic solutions of cyanide containing solutions.
Additional support of this latter point as well as the overall effect of oxides on the surface was given in a subsequent study
by Chen et al. [91], which examined the SERS and SH response
of silver electrodes in sulfate- and cyanide-containing solutions.

The SH response from silver electrodes cycled in similar
electrolytes has also been compared with other optical measurements of the silver surface in an effort to understand the various chemical species and enhancement mechanisms responsible for
the SH response during various stages of surface oxidation and
reduction. Several studies have noted the similarities and differences in the potential dependence of the broadband lumines-
cence which accompanies the SH response [89,90,76]. Marshall
and Korenowski provided evidence for the existence of AgCl–Ag
cluster complexes which they believe play a role in the overall
signals produced [89,90]. Optical resonances between 2ω and
electronic transitions in these complexes have been proposed as
a mechanism for the enhancement. Photoacoustic (PA) measurements of these silver/electrolyte systems performed by Richmond
and Chu [76,93] demonstrated similar optical resonances between
the fundamental and harmonic optical fields and the oxidatively
formed surface species. In these experiments, the optical ab-
sorptivity of the surface as measured by the PA method was
monitored at 1060 nm and 532 nm during an ORC in several
electrolytes and the results were compared with the SH response.
A direct correlation between the strength of the photoacoustic
response and the SH enhancement was noted. The combination
of the two techniques provided insight into surface species present during the ORC as well as possible mechanisms for SH enhancement.

   The fact that these initial efforts focused almost exclusively
on silver does not suggest that surface SHG can only be applied
to this metal.   In fact a large variety of solid surfaces produce
measurable signal levels with and without the presence of roughness features.   Figure 7 demonstrates the general applicability
of SHG to a variety of materials under both conditions.   This

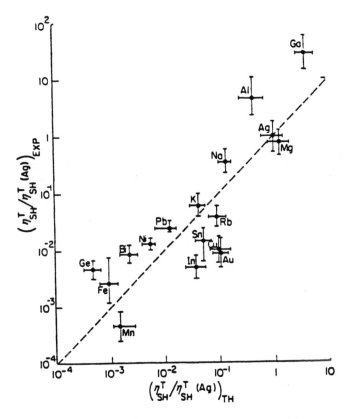

FIG. 7.   Comparison of experimental data ($\cdot$) and theoretical
calculation (---) of the local field enhancement factor, η_{SH}^T,
for various metals and Ge relative to that of silver, $\eta_{SH}(Ag)$.
(From Ref. 148.)

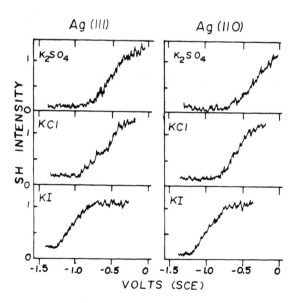

FIG. 8. SH intensity at 532 nm as a function of applied voltage (vs. SCE) for Ag(111) and Ag(110) electrodes biased in the indicated 0.1 M aqueous electrolyte solutions. All scans are on the same relative scale. Scan rate = 5 mV/sec. (From Ref. 71.)

work by Boyd et al. [148] was performed to study the local-field enhancement due to surface roughness on various materials ranging from the alkalis to a semiconductor. The ratio of the signal from the roughened and smooth surfaces was used to determine the enhancement factor. With laser excitation at 1064 nm, the observed second harmonic enhancements for different materials varied from 27 to 1×10^{-3} times that of silver. A simple model was proposed for estimating the enhancement factor for various materials.

In recent years, the SH response during an oxidation-reduction cycle has been examined for a variety of metals. Polycrystalline copper and gold have been examined by Richmond [149]. For copper, the formation of the halide complexes on the surface leads to a large increase in the SH response at 532 nm. In contrast, the SH response from the gold surface decreases as the surface is being oxidized to form gold–halide complexes rather than showing a large signal increase.

Biwer et al. [81] investigated the aqueous corrosion of iron electrodes with particular attention to the passive oxide layer that forms on this electrode in basic solutions. In contrast to previous studies with nanosecond lasers, a frequency doubled and Q-switched picosecond laser operating at 1.7 kHz was employed. Since iron has one of the weakest SH responses of the 17 metals examined by Boyd et al. [148], it offers a particularly interesting test case for the use of the picosecond laser system. The SH response from this electrode during cycling was found to be readily observable with photon counting electronics. The optical measurements were found to correlate directly with the electrochemical current-voltage measurements and to provide complementary information. At least two intermediates in the reduction of the passive layer were observed in both borate buffer (pH 8.5) and in more basic NaOH solutions. Compositional changes were observed as large changes in SHG signal levels in potential regions where no faradaic oxidation changes in the electrode surface occurred as seen by the featureless cyclic voltammogram. A later study by this group examined the corrosion of nickel in alkaline solutions [98]. Both 1064 nm and 532 nm probe wavelengths were used here to deconvolute the various oxide species on the surface.

B. Ionic and Molecular Adsorption

Examination of the sensitivity of SHG to nonfaradaic processes such as the adsorption of ions and molecules at the metal/liquid interface has been the focus of many studies over the past few years. Even with these studies, deriving quantitative parameters for adsorption from the SH response remains an elusive goal.

Nevertheless, considerable progress has been made in understanding the fundamental parameters responsible for the potential dependent SH response when the electrode is polarized within
the IPR.

When a silver electrode surface is polarized within the limits
of solvent reduction and metal oxidation (commonly referred to
as either the ideally polarizable region or double layer charging
region), a potential dependence in the SH response is observed.
The first report of this effect appeared in 1967 [13]. They
attributed the potential dependence to the presence of the applied electric field, E_{dc}, to the surface which adds an additional
factor of:

$$\overline{P}^{NLS}(2\omega) = \chi^{(3)} : \overline{E}(\omega)\overline{E}(\omega)\overline{E}_{dc} \tag{8}$$

to the nonlinear polarizability expression in Eq. (1). Unlike
the second-order process of SHG, this third-order term is not
restricted to regions of broken symmetry. However, the localization of the electric field to the surface of the metal isolates
this additional contribution to the surface also. Furthermore,
even though the third-order susceptibility, $\chi^{(3)}$, is several
orders of magnitude smaller than $\chi^{(2)}$, the contribution from
this effect can be significant due to the magnitude of $\overline{E}_{dc}$, which
at an electrochemical interface can be as large as 10^6 V/cm. In
this initial study, the observed potential dependence was also
found to vary with input polarization and wavelength. The two
fundamental wavelengths used were 8658 Å and 5632 Å. One
problem plaguing these early studies, as well as most of the surface SH work done in the 1960s and early 1970s, was the condition of the surface, i.e., roughness or presence of oxides.

In 1984, two sets of studies appeared which examined this
phenomenon in more detail. Both investigated the SH response
from silver surfaces biased in simple electrolytes within the

ideally polarizable region. Corn et al. [61,103] examined the
surface plasmon-enhanced SH response from silver films potentio-
stated in perchlorate-containing solutions. Richmond studied
the SH response from silver polycrystalline and single crystal
surfaces biased in a variety of adsorbing and nonspecifically adsorbing electrolytes [70–72]. Both noted a measurable SH response was observable when nonspecifically adsorbing anions
such as perchlorate and sulfate ions were at the surface. Since
the hydrated water molecules surrounding these ions as well as
the silver surface would only make a negligible contribution to
the SH response relative to the silver surface, it was presumed
that the change in the SH response reflected the surface rather
than the adsorbate properties. In addition, Richmond showed
that the magnitude of the SH response during potential bias
showed only a negligible dependence on the type of anion adsorbing on the surface. For example, the larger optical polarizability
of I^- was not reflected in a larger SH response relative to the
lesser polarizable Cl^- or Br^- ions at comparable surface concentrations. Instead, the ions gave a comparable SH response at
the most adsorbing potentials. The onset of the potential induced rise in the SH response from the various anions and single
crystal faces was found to vary in correspondence with the PZC
(Fig. 9). Whereas all systems showed a rise in SHG with increased adsorption of anions, a shift in this rise to more negative potentials was observed with increased absorptive character
of the different ions. The primary conclusion drawn from this
study was that the SH response from the interface was not responding directly to the ions but was instead reflecting the
changing optical properties of the electrified surface as it was
charged and discharged.

Corn et al. [103] proposed a quantitative model for the observed behavior from these studies. Using Eq. (8) and assuming that Gauss's Law was appropriate for the surface for which

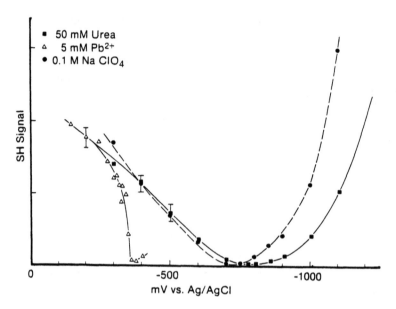

FIG. 9.   SHG at 532 nm from a polycrystalline silver film (45
nm) in contact with a solution of 0.1 M NaClO$_4$ (solid circles)
as a function of electrode potential (versus Ag/AgCl).   Also
shown are SHG potential curves for a solution of 50 mM urea +
0.1 M NaClO$_4$ (solid squares) for a solution of 0.1 M sodium
acetate + 5 mM lead acetate (open triangles) at a polycrystalline
silver film.   (From Ref. 61.)

$\overline{E}_{dc} = 4\pi\sigma$, their model predicts that the SH response should be
parabolic around the PZC, scaling quadratically with the surface
charge density σ.   The results of their surface plasmon-enhanced
experiments in perchlorate-containing solutions are shown in Fig.
10.   The model shows a good correlation with the data.   In a
later study [103], they examined the validity of the surface
charge density model in more detail.   Using a silver-mica capacitor as an experimental model for the double layer, they found
that the SH response correlated well with the charging of the
capacitor.   The surface charge density model was in qualitative

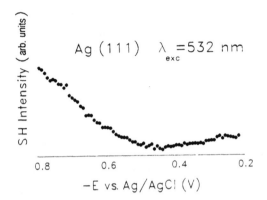

FIG. 10.   SH intensity as a function of voltage (vs. Ag/AgCl) within the IPR for a Ag(111) crystalline 0.25 M Na_2SO_4.   P-polarized incident wavelength of 532 nm was used with the p-polarized SH intensity collected at 266 nm.   The data was taken at an azimuthal angle of 30° relative to the [2$\bar{1}\bar{1}$] direction of the crystal.   (From Ref. 101b.)

agreement with the related electrochemical measurements but deviations were clearly evident.   They attributed these deviations to the assumption in the model that the charge on the electrode varies linearly with applied potential.   A second reason noted for the deviation was the much higher electric fields obtained in the electrochemical cell which would require inclusion of higher order terms than the linear hyperpolarizability used in their model.

Rojhantalab and Richmond [73] examined the correlation between the SH response for the polycrystalline silver electrode and the surface charge density model with the use of differential capacitance.   They found a good correlation for nonspecifically adsorbing ions at lower coverages where $\sigma < 40$ $\mu C/cm^2$.   However, a significant deviation from the model was found for higher concentrations.   The deviation was even more apparent for specifically adsorbing anions.

The effect of surface roughness on these measurements in the IPR has also been examined by Richmond [92] with some interesting effects observed. First, as with other studies of silver surfaces, enhancement in the SH response was observed after the electrode had been subjected to slight electrochemical roughening. Second, the potential dependence became parabolic and approached a similar form to that obtained in Fig. 9 for the surface plasmon-enhanced silver films of Corn et al. [61,103]. Differential capacitance [73] measurements showed that this change in the SH response was not merely reflective of a higher surface charge density in the cation adsorption region but was due to a surface optical effect. Simultaneous photoacoustic measurements [76,93] also showed an enhanced optical coupling to the surface when it was roughened. The potential dependence in the acoustic response directly correlated with the potential dependence in the SH response from the smooth and roughened surface. Based on these results, the similarity of the SH potential dependences observed in the roughening studies and the surface plasmon work, and the fact that surface plasmons can be accessed when the originally smooth surface is roughened, the authors proposed that the roughening induced parabolic potential dependence merely reflected the potential dependence of the surface plasmons.

A recent report by Tadjeddine et al. [150] provides additional information to answer questions regarding the differing potential dependences observed under the various conditions. This study reports that the potential dependence that one obtains from these surfaces is highly dependent on the angle of incidence. For incident angles close to 45° where the Richmond work was performed, one obtains a potential dependence shown in Fig. 8. When an angle approaching 75° is used, the parabolic potential dependence is observed. They attribute this change in potential dependence with incident angle to the angular dependence of the Fresnel coefficients. This new angular dependent

data might help to partially explain the observed effect of surface roughening on the SH response. Since the incident angle
is not well defined on these roughened surfaces but is a mix of
different angles, the resulting SH response would then be a sum
of the various incident optical/surface geometries.

The above model for describing the SH response from the
silver electrode surface examined with 1064 nm light assumes a
free electron gas type response from the electrons. Unfortunately this is only an approximation as will be evident in the
description of the rotational anisotropy results for silver shown
in Sec. IV.C. In these studies, the assumption of a free electron gas for silver at 1064 nm is not consistent with the observed anisotropy in the SH response from the crystalline silver
electrode surfaces examined. The effect of interband transitions
on the potential dependence of the SH response in the double
layer region has been examined in more detail in studies by
Georgiadis et al. [101a,b]. Using a 532 nm probe beam on a
crystalline silver surface and detecting the SH response at 266
nm, the potential dependence is found to be completely opposite
to those results obtained at 1064 nm. Figure 10 shows the results obtained for experiments run under similar conditions as
those in Fig. 8 and with Na_2SO_4 as the supporting electrolyte.
A maximum in the SH intensity is observed at the PZC followed
by a rapid decline as the surface is charged positively. At
these wavelengths, an optical resonance with either an interband
transition or surface state is plausible. At 1.064 μm excitation,
both the fundamental and the second harmonic photon energies
are well below the silver interband transitions (3.8 eV), while
at 532 nm excitation, the SH photon energy is 4.66 eV, well
above the interband transition energy. The authors have attributed this difference to a resonant coupling to surface states
which can be altered by the variation of the large dc electric
field at the surface. This will be described in more detail in
Sec. IV.C.

Time-resolved measurements have also been used to evaluate
the dependence of the SH response on the surface charge density of silver at the longer wavelength. This was first performed by Corn et al. [61]. In more recent experiments by
Robinson and Richmond [95] in which the kinetics of adsorption
and desorption of anions on the surface were explored in more
detail, a fast potential step within the IPR was applied to a
polycrystalline silver electrode surface in Na_2SO_4. The SH response was then interrogated immediately and "continuously"
following the step with the use of the fundamental of a mode-
locked Nd:YAG laser operating at 76 MHz. This allowed a better
continuous probe following the potential step than was possible
with the earlier study using a 10 Hz Nd:YAG laser. The transient SH response to a fast step between the -0.65 V and -0.1
V is shown in Fig. 11a and for the reverse step in Fig. 11b.
The corresponding cell current transients decayed via a single
exponential, as expected for simple charging processes in the
absence of electron transfer. The response time of the cell was
determined to be 13.8 ± 0.8 msec. The solid curves in the figure are the SH transients computed by a model in which the
time-dependent surface charge density is incorporated. The
expressions are given in the figure. With further detailed analysis of the transients the authors concluded that the time-
dependent SH data correlate with the time-dependent surface
charge density in a manner not consistent with the surface
charge density model of Corn et al. [61]. Particularly for anion
adsorption, the second harmonic signal shows the closest linear
correlation with the charge density on the electrode.

Molecular absorption has also been examined in a number of
studies. As noted earlier for roughened silver surfaces, an increase in signal was observed in the SH response at negative
potentials when pyridine was added (Fig. 5) [83]. In the presence of the adsorbed pyridine, the SH response was found to

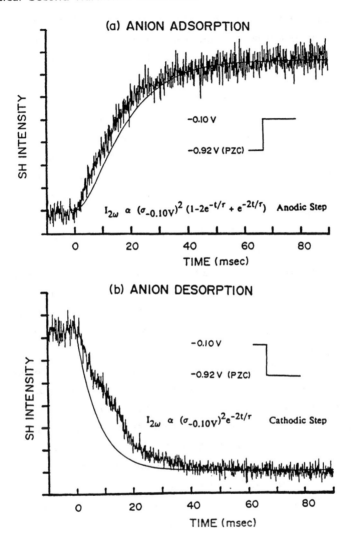

FIG. 11. Transient SH response from a polycrystalline Ag electrode to a potential step within the IPR. 1.06 μm light from a
cw mode-locked Nd:YAG laser was focused on the surface. (a)
SH response to a step between the PZC and −0.10 V and (b)
SH response to the reverse step in 0.25 M Na_2SO_4. The solid
curves are calculated fits to the data using a cell time constant
of $\tau \simeq 14$ msec. (From Ref. 95.)

increase with potential bias negative to the PZC. The authors
attributed this potential dependence to a change in the molecular
orientation of the pyridine relative to the silver surface. Later
studies suggested that this change in SH response with negative
bias was not due to the adsorbed pyridine explicitly but was reflecting an enhanced coupling of incoming and harmonic fields to
the roughened silver surface due to the presence of pyridine
[71]. This conclusion was based on studies showing that the
SH response from pyridine on a surface in the absence of roughening does not show the type of potential dependence observed
for the roughened surface. Combined differential capacitance
and SHG studies on smooth and roughened surface suggest that
although the optical nonlinearity of the adsorbed pyridine itself
is likely contributing, the dominate contribution to the increased
SH response observed at potentials where pyridine is adsorbed
comes from the surface itself, which is altered by the presence
of pyridine [73]. Pthalazine has also been examined on silver
surfaces by SHG [77].

  For the above studies on silver, excursion into the region
of solvent reduction has been avoided due to the large fall-off
in SH intensity as hydrogen evolution occurs [71]. Campbell
and Corn [78] have examined the effect of presence of hydrogen
evolution in the SH response on silver electrodes in more detail
in acetonitrile solutions containing millimolar concentrations of
various acids. Since it is unlikely that the adsorbed hydrogen
moiety contributes to the nonlinear susceptibility, the observed
modifications in the SH signal must result from changes in the
optical response of the electrons at the metal surface. For a
series of acids studied, they found that the only reaction mechanism which correctly predicts the potential and hydrogen activity
dependence of the SH signal was one in which the creation of
molecular hydrogen proceeds by a two-step process: the formation of adsorbed monatomic hydrogen followed by a surface combination reaction of two adsorbed hydrogen atom species.

The potential dependence of the SH response from polycrystalline platinum electrodes has been the focus of other studies
by Campbell and Corn [94]. In their aqueous sulfuric and
perchloric acid solutions, the SH signal is found to increase during the chemisorption of strongly and weakly bound hydrogen
species onto the platinum surface. In these studies, a Nd:YAG
pumped dye laser was used to generate the fundamental 578 nm
beam. Figure 12 shows the SH response during a cyclic voltammogram from a polycrystalline platinum electrode in sulfuric acid.
The SH signal reaches a minimum near the PZC at ~0.100 V and
rises steadily until +0.500 V. They attribute this rise in the
double layer charging region to bisulfate adsorption. The SH
signal then drops with positive sweep as surface oxides form.
They attribute the rise in SH signal at potentials cathodic of
the PZC to adsorption of hydrogen atoms on the platinum surface which occurs prior to H_2 evolution. In the presence of
iodine, the hydrogen adsorption is blocked, which leads to a
diminished SH response in this region. A comparison of the SH
signal with the amount of charge passed during the hydrogen
deposition reaction reveals that the nonlinear susceptibility of
the surface is linearly related to the surface coverage of the adsorbed hydride species.

All of the above studies of adsorption have been performed
under conditions in which the fundamental or SH beam is of an
energy far removed from any optical resonance with the adsorbing species. In the first demonstration of the use of resonant
molecular SHG from an electrochemical interface, Campbell et al.
[97] have studied methylene blue chemisorbed on a sulfur-
modified polycrystalline platinum electrode. The sulfur monolayer prevents the irreversible decomposition of the dye molecules on the platinum electrode and ensures that the contributions to the surface nonlinear susceptibility from the metal and
from the methylene blue remain separable. The fundamental

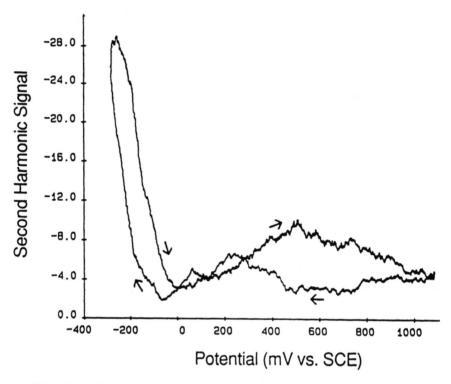

FIG. 12. The second harmonic signal (arbitrary units) during a potential cycle (vs. Ag/AgCl) from a polycrystalline platinum electrode in a 0.5 M H_2SO_4 solution. Scan rate: 10 mV/sec. Incident wavelength is 578 nm. (From Ref. 79.)

wavelength used was 585 nm with the second harmonic photons collected at 292.5 nm. With these wavelengths, which are both coincident with absorption bands for methylene blue, the SH response is enhanced by both the incoming and outgoing fields. Figure 13a and b show the current and SH response, respectively, for the platinum electrode cycled in sulfur-containing electrolyte. Both are relatively featureless. In Fig. 13c, both the CV curve and the SH response are depicted for this electrode in the presence of methylene blue. The measured potential

dependence of the resonant SH signal shown in this latter figure matches that expected during the electrochemical reduction
and reoxidation of the chemisorbed methylene blue. Polarization
studies were used to provide information about the average molecular orientation of the dye molecules as a function of coverage.

C. Structural Properties of Single Crystal
Electrodes In Situ

A wide variety of techniques are available for examining the
morphology of the surface and its electronic structure in UHV.
As a result, many different metallic and semiconductor surfaces
have been characterized with respect to such properties as electronic surface states, surface stability and reconstruction, and
adsorbate interactions. In the last few years techniques have
been developed which have the potential to enable electrochemists
to make similar in situ measurements of such properties. Prior
to this time such information has been derived from ex situ analysis by either simulating the electrochemical environment in UHV
or by performing round-trip excursions between the two environments and comparing the surface characteristics in and out of
solution. Two in situ techniques which have begun to show particular promise are x-ray scattering [151,152] and scanning tunneling microscopy [153,154]. Recent studies have suggested
that SHG may also be a player in this area.

The SH studies described thus far have taken advantage of
the localization of the SH response to the interfacial region as a
means of surface examination. Yet to be discussed is the use of
the rotational anisotropy in the SH response, which results from
the interaction with the incident beam and a single crystal surface as it is rotated about its surface normal. This approach
has the potential to provide valuable information about surface
geometrical structure and the degree of atomic ordering of the
surface in situ. Although measurements of surface ordering and

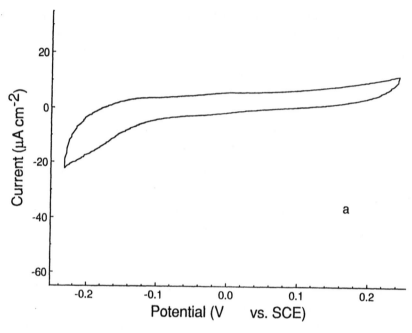

FIG. 13. The cyclic voltammogram (a) and nonresonant SHG
signal at 295.2 nm (b) from a polycrystalline platinum electrode
in a 1 mM Na_2S + 0.5 M H_2SO_4 solution. (c) The potential dependence of the in situ resonant SHG signal at 292.5 nm from a
monolayer of methylene blue chemisorbed onto a sulfur-modified
platinum electrode during reduction of the monolayer. The
monolayer is chemisorbed onto the sulfur-modified platinum electrode in a phosphate buffer solution at pH 7.9. The SHG signal
was obtained from a p-polarized fundamental 585 nm beam at an
incident angle of 40.5°. All scans are at 10 mV/sec. (From
Ref. 97.)

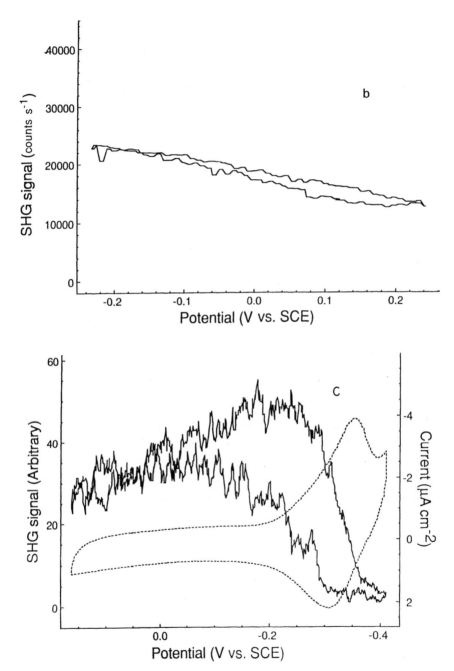

FIG. 13 (Continued).

symmetry made by this method are macroscopic in nature, the time-resolving capabilities of the technique add an additional dimension that is not available with any other surface probe. Furthermore, wavelength-dependent studies have the potential of providing valuable information about surface electronic properties and surface states, a nearly unexplored area in electrochemistry.

Most of the current studies where SHG has been used in this manner, particularly in the electrochemical cell, have been exploratory in nature. To place the electrochemical studies in perspective, the discussion will begin with a brief review of some of the principles behind the polarization dependence of the SH response from crystalline surfaces incorporating some of the pioneering work done in this area in UHV and under ambient conditions. This will be followed by a description of studies performed on native metal surfaces in solution. The next section on metal deposition will discuss the extension of these studies to the growth of metal overlayers on these single crystal surfaces.

1. Rotational Anisotropy from Single Crystal Surfaces

The first successful studies showing that the SH response from a centrosymmetric surface varies as the crystal is rotated about the surface normal were reported by Guidotti, Driscoll, and Gerritsen in 1983 for Si(111) and Ge(111) under ambient conditions [155]. Tom, Heinz, and Shen [156] soon thereafter reported SH rotational anisotropy from etched Si crystals in air but at lower pulse energies, which allowed them to perform a more complete study of these surfaces without the interfering effects of surface breakdown which were present in the earlier studies. Figure 14a shows the variation in the s-polarized SH intensity measured for p-polarization excitation, $I(\parallel, \perp)$, as a Si(111) sample was rotated about the surface normal. $I(\parallel, \perp)$ refers to $p(\parallel)$ and $s(\perp)$ polarized (fundamental, SH) light.

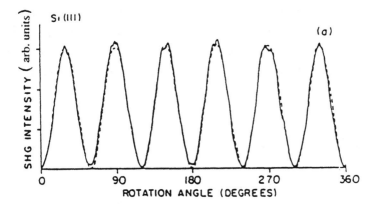

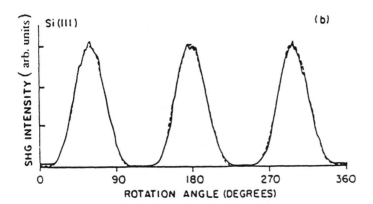

FIG. 14.   SH rotational anisotropy from a Si(111) surface in air
under p-polarized 532 nm excitation.   Excitation beam from a Q-
switched Nd:YAG laser incident at 45° from the surface normal.
The rotation angle lies between the x axis in the plane of incidence and the [100] axis projection on the surface.   (a) S-
polarized SH data (solid line) and theoretical fit (dashed line)
to Eq. (12) (with A = 0).   (b) P-polarized SH data and theoretical fit to Eq. (12) (with B = −A).   (From Ref. 160; also shown
in Ref. 156 with a different definition of the angle of rotation.)

Anisotropic patterns with threefold and fourfold symmetry were
observed for $I(\parallel, \parallel)$ for Si(111) (Fig. 14b) and Si(100), respectively [156]. The former result agrees with Driscoll and Guidotti
[155], although these workers had earlier found both $I(\parallel, \parallel)$ and
$I(\perp, \parallel)$ to be isotropic for Si(100) [155].

Tom et al. [156] and later Sipe et al. [161] proposed similar
theories to describe the observed anisotropic response from these
crystalline surfaces. Both note that the symmetry of the crystal
and its surface governs the observed anisotropy and that the
nonlinear response arises from the surface electric dipole, and
bulk electric quadrupole and magnetic dipole contributions in
the crystal. These models are an extension of the macroscopic
theory of Shen and co-workers described in Sec. II, to the special case of cubic media under single plane wave excitation. The
model proposed by Tom et al. [156] ascribes the effective nonlinear polarization, $\bar{P}_{s,eff}(2\omega)$ (Eq. (5)), to a linear combination
of the bulk nonlinear polarization, $\bar{P}(2\omega)$ (Eq. (3)), and a surface nonlinear polarization, $\bar{P}_S(2\omega)$ (Eq. (1)). For a cubic medium excited by a single beam at frequency ω, the bulk nonlinear
polarization induced by $\bar{E}(\omega)$ becomes:

$$P_i(2\omega) = \gamma\nabla_i[\bar{E}(\omega) \cdot \bar{E}(\omega)] + \zeta E_i(\omega)\nabla_i E_i(\omega) \qquad (9)$$

where γ and ζ are frequency-dependent material parameters and
i refers to the crystal principle axes. These two terms arise
from the induced quadrupole and magnetic dipole moments in the
bulk. The first term is independent of crystal orientation, and
it is the second term which leads to anisotropy.

The effective surface dipole polarization density can be written as:

$$P_i(2\omega) = \chi^{(2)}_{ijk}:E_j E_k \delta(z - z_0^+) \qquad (10)$$

where the surface susceptibilities are in terms of beam coordinate
geometries (s,k,z). The form of $\chi^{(2)}$ in surface coordinates

(x,y,z) is obtained by assuming perfect termination of the bulk
symmetry. The nonzero and equivalent tensor elements are determined by noting that the form of the tensor must remain in-
variant under symmetry operations which transform the material
into itself [157]. Transforming $\chi^{(2)}$ from surface to beam coordinates is then a straightforward application of the transforma-
tion law of a third rank tensor:

$$T'_{ijk} = a_{il} a_{jm} a_{kn} T_{lmn} \qquad (11)$$

where a_{ij} is an element of the direction cosine matrix. In this
manner, $P_i(2\omega)$ is given in terms of the surface susceptibility
elements $\chi_{ijk}(x,y,z)$. In the case of the (111) surface, which
has C_{3v} (3m) symmetry, the nonzero elements are: χ_{zzz},
$\chi_{zxx} = \chi_{zyy}$, $\chi_{zzx} = \chi_{xxz} = \chi_{yyz} = \chi_{yzy}$, and $\chi_{xxx} = -\chi_{yyz} =$
$-\chi_{xyy} = -\chi_{yxy}$, where the z direction is perpendicular to the
surface and the mirror plane lies parallel to the x direction along
the [2$\bar{1}\bar{1}$] crystal direction (Fig. 15). The first three independent elements have a field component perpendicular to the surface, while χ_{xxx} involves fields only in the plane of the surface.

The structural symmetry of the surface is reflected in the
form and magnitude of the tensor elements of the surface nonlinear susceptibility, $\chi^{(2)}{}_s$, and the bulk anisotropic susceptibility, ζ. The topmost atomic layer of a perfectly terminated
(111) crystal has 6m symmetry. Inclusion of the next atomic
layer reduces the symmetry to 3m. For 1.06 µm excitation, the
penetration depth of $\bar{E}(\omega)$ is about 100 Å. The electric quadrupole allowed contribution to $\bar{E}(2\omega)$ from the decaying incident
field is attenuated by e^{-2} relative to the surface dipole contribution. The surface electric dipole contribution is thought to
arise from the first 10 Å [158]. Consequently, the reduction
of the symmetry to 3m is physically reasonable. General solutions for the s- and p-polarized SH electric fields using Eqs.
(9) and (10) are given in Ref. 156. The expressions for a

(a)

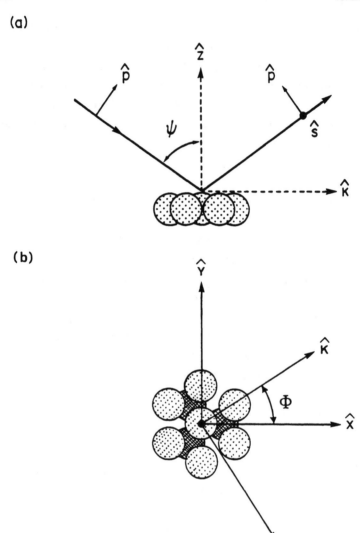

(b)

FIG. 15. Diagrammatic representation of the experimental crystal and beam geometries for p-polarized radiation incident upon the (111) crystal face as viewed (a) from the side and (b) from the top including the second atomic plane. The crystal coordinates are labeled x,y,z with the x direction along the [2$\bar{1}\bar{1}$] crystal direction. The beam coordinates are labeled s,k,z. (From Ref. 62.)

given input and output polarization geometry explicitly display
the appropriate symmetry for the Si(111) crystal results displayed
in Fig. 16a and 16b for the appropriate optical input and output
polarizations.

Rotationally anisotropic SH measurements since Litwin's work
[159] are generally discussed in light of Tom's model [156,160].
The model predicts a simple functional dependence of the SH
intensity on the angle of rotation of the sample about the surface normal:

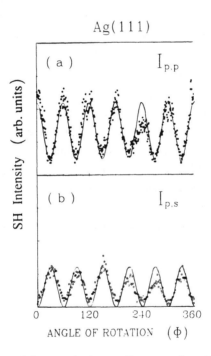

Ag(111)

FIG. 16. Second harmonic intensity as a function of angle of
rotation for Ag(111) immersed in 0.25 M NaClO$_4$, pH = 5.0 at
−0.72 V (PZC) with p-polarized 1064 nm illumination. Angle of
incidence (ψ) is 10°. The solid lines are fits to the data generated from the theoretical expressions given in the text (Eqs. (13)
and (15)). (a) P-polarized SH intensity, a/c$^{(3)}$ = 0.54i. (b) S-
polarized SH intensity. The constant b$^{(3)}$ is taken to be unity.
(From Ref. 62.)

$$I(2\omega) = \mid A + B \, f(\psi) \mid^2 \tag{12}$$

The definition of ψ is arbitrary and varies between researchers; it is usually taken as the angle between an axis in the laboratory frame, such as parallel or perpendicular to the plane of incidence, and a mirror plane or crystal axis direction in the surface. The function $f(\psi)$ reflects the 2mm, 3m, and 4mm symmetry of the (110), (111), and (100) surfaces, respectively. The constants A and B represent the respective isotropic and anisotropic contributions to the SH intensity which depend on the crystal, the experimental geometry and the fundamental and SH polarizations. For example, a fit to the data in Fig. 16 using Eq. (12) yields $A = 0$ and $f(\psi) = \sin 3\psi$ for $I(\parallel, \perp)$, whereas $\mid A/B \mid = 1$ and $f(\psi) = \cos 3\psi$ for $I(\parallel, \parallel)$.

The functional form of the rotational anisotropy for cubic centrosymmetric media has also been predicted by a theory developed by Sipe, Moss, and van Driel [161]. As in the work of Tom et al. [156,160], no assumption is made about the microscopic origin of the bulk and surface nonlinear polarization. The theory assumes a macroscopic surface and bulk symmetry for a given crystal orientation and considers possible electric dipole, electric quadrupole, and magnetic dipole sources for the nonlinear response. They derive expressions for the total reflected p- and s-polarized SH fields from perfectly terminated (111), (100), and (110) crystals under p- and s-polarized excitation. For example, SH fields for the first two crystal faces are:

$$E^{(2\omega)}(\parallel, \parallel) \propto a_{\parallel, \parallel} + c^{(m)}_{\parallel, \parallel}\cos(m\phi), \tag{13}$$

$$E^{(2\omega)}(\perp, \parallel) \propto a_{\perp, \parallel} + c^{(m)}_{\perp, \parallel}\cos(m\phi), \tag{14}$$

$$E^{(2\omega)}(\parallel, \perp) \propto b^{(m)}_{\parallel, \perp}\sin(m\phi), \tag{15}$$

$$E^{(2\omega)}(\perp, \perp) \propto b^{(m)}_{\perp, \perp}\sin(m\phi), \tag{16}$$

For the (111) and (100) surfaces, m = 3 and m = 4, respectively.
The isotropic coefficient a and the anisotropic coefficients $b^{(m)}$
and $c^{(m)}$ can have both bulk and surface contributions and depend on the crystal symmetry.   The linear and nonlinear dielectric constants of the material, as well as the appropriate Fresnel
factors at ω and 2ω, are incorporated into the constants $a, b^{(m)}$,
and $c^{(m)}$.   The susceptibilities contained in each of these constants are given in Table 1.   The term $b^{(m)}$ is linearly proportional to $c^{(m)}$ where the constant of proportionality is determined
by the dielectric constants of the media at ω and 2ω as well as
the angle of incidence through the Fresnel coefficients.   The
azimuthal angle ϕ is defined as the angle between the [2$\bar{1}\bar{1}$] direction and the component of the incident wavevector parallel to
the surface (Fig. 15).   Explicit expressions for $a, b^{(m)}$, and
$c^{(m)}$ for a given crystal face and polarization geometry are given
in Ref. 161.   The forms predicted by Tom's model [156] and
Sipe's model [161] are equivalent, given that $\psi = \phi$, although
the coefficients $a, b^{(m)}$, and $c^{(m)}$ are defined differently in each
model.   The relationship between the coefficients A and B in Eq.
(12) and the coefficients $a^{(m)}, b^{(m)}$, and $c^{(m)}$ from the theory
of Sipe et al. [161] (Eqs. (13) to (16)) is apparent by inspection.   This phenomenological theory also describes the SH rotational anisotropy data cited above.   It is apparent from Eq. (13)
combined with Table 1 that χ_{zzz}, χ_{zxx}, χ_{xzx}, and γ contained
in the a term do not show any dependence on the azimuthal
angle ϕ and therefore are referred to as isotropic susceptibilities.
Thus, two physical parameters which can be separated and monitored experimentally are the isotropic susceptibilities contained
in a and the anisotropic susceptibilities contained in b.   This
can be accomplished by following $I^{(2\omega)}(\parallel, \parallel)$ and $I^{(2\omega)}(\perp, \parallel)$ at
an azimuthal angle of $\phi = 30°$.

Similarly, the nonvanishing tensor elements for the (110)
face which has C_{2v} (mm2) symmetry are: χ_{zzz}, χ_{zxx}, χ_{zyy},

TABLE 1

Composition of the Parameters a, b, and c in Terms of the Surface and Bulk Susceptibility Elements. P-polarized (♦) or s-polarized (◊) Fundamental Field. i = (x,y).

(111) 3*m*

	χ_{zzz}	χ_{zii}	χ_{izi}	χ_{xxx}	ζ	γ
a	♦	♦,◊	♦		♦,◊	♦,◊
$c^{(3)}$				♦,◊	♦,◊	
$b^{(3)}$				♦,◊	♦,◊	

(110) *mm*2

	χ_{zzz}	χ_{zxx}	χ_{zyy}	χ_{xzx}	χ_{yzy}	ζ	γ
a	♦	♦,◊	♦,◊	♦	♦	♦,◊	♦,◊
$c^{(2)}$		♦,◊	♦,◊	♦	♦	♦,◊	
$c^{(4)}$						♦,◊	
$b^{(2)}$				♦	♦	♦,◊	
$b^{(4)}$						♦,◊	

(100) 4*mm*

	χ_{zzz}	χ_{zxx}	χ_{xzx}	ζ	γ
a	♦	♦,◊	♦	♦,◊	♦,◊
$c^{(4)}$			♦,◊	♦,◊	
$b^{(4)}$			♦,◊	♦,◊	

$\chi_{xzx} = \chi_{xxz}$, and $\chi_{yzy} = \chi_{yyz}$, where the directions x and y are along the (110) and (001) axes, respectively. The p- and s-polarized SH intensities are then given as:

$$I^{(2\omega)}(\|,\|) \propto [a + c^{(2)} \cos(2\phi) + c^{(4)} \cos(4\phi)]^2 \qquad (17)$$

$$I^{(2\omega)}(\|,\perp) \propto [b^{(2)} \sin(2\phi) + b^{(4)} \sin(4\phi)]^2 \qquad (18)$$

respectively. The anisotropic and isotropic surface and bulk susceptibilities contained in the constants for the (110) face are also listed in Table 1, where it is shown that the terms $b^{(4)}$ and $c^{(4)}$ are linearly proportional. There are two tensor elements contained in $c^{(2)}$, χ_{zzz}, and χ_{zyy}, which do not appear in $b^{(2)}$. For this reason an additional anisotropy may appear in the p-polarized SH response which is not evidenced in the s-polarized signal for the 110 crystal. The angle ϕ, in this case, is defined as the angle between the [110] direction and the component of the incident wavevector parallel to the surface. The dielectric constants of the material and the appropriate Fresnel factors at ω and 2ω are incorporated into the constants a, $b^{(2)}$, $b^{(4)}$, $c^{(2)}$, and $c^{(4)}$. By orienting the 110 face at an azimuthal angle of 45°, a and $(b^{(2)} + b^{(4)})$ can be separately monitored by measuring $I^{(2\omega)}{}_{p,p}$ and $I^{(2\omega)}{}_{p,s}$, respectively.

2. DC Electric Field Effects

When a dc electric field is applied to the interface such as in the electrochemical cell, an additional contribution to the nonlinear polarizability of the material will be induced through the third-order susceptibility $\chi^{(3)}$ [13]. As noted earlier, this contribution can be comparable in magnitude to the second-order response when the dc field is large as is the case for a metal electrode in solution held at a bias potential positive of the potential of zero charge. The SH fields can be viewed as arising from a sum frequency mixing process where two incident ac fields at frequency ω are mixed with a dc field at frequency $\omega = 0$ to produce a reflected and transmitted wave at frequency 2ω. $\chi^{(3)}$ has the properties of a fourth-rank tensor so that the third-order source polarization has the form:

$$P_i(2\omega) = \chi^{(3)}{}_{ijkl} : E_j E_k E_l \tag{19}$$

where $E_l = E_{dc}$. Making the appropriate transformations from surface to beam coordinates, it can be shown that $\chi^{(3)}{}_{ijkl}$ has the same from as the second-order surface tensor $\chi^{(2)}{}_{ijk}$. This is a result of the applied field being parallel to the surface normal about which the sample is rotated. Because the field is localized at the surface, the $\chi^{(3)}$ effect should mimic the dipolar surface response and an effective susceptibility can be written as:

$$\chi^{(2)}_{eff} = \chi^{(2)}_{ijk} + \chi^{(3)}_{ijkl} E_{1(\omega=0)} \tag{20}$$

The result is that the application of an applied field would not be expected to change the form of the source polarization, and consequently the symmetry observed in the SH rotational anisotropy from the interface should remain unchanged. The magnitude of the individual tensor elements in $\chi^{(2)}$ and $\chi^{(3)}$ may, however, differ appreciably leading to a change in the observed SH patterns.

3. Electrochemical Applications

All of the single crystal electrode work to date has been on metallic surfaces and particularly the noble metals. As in the UHV surface studies in this area, the work has generally been exploratory in nature. The goal of this early work on native surfaces has been to understand how the optical observations correlate with surface geometrical and electronic structural properties. The first step in achieving such a goal is to understand the source of the nonlinear polarizability from the metallic surface in either a solution or UHV environment. At present this has primarily meant correlating the observations with the existing macroscopic theories [160,156,161] described earlier in this section in attempts to understand the relative magnitude of the tensor elements, both surface and bulk, as well as monitoring the sensitivity of such elements to surface alterations. The advantage of making such exploratory studies in the electrochemical cell is the ease with which surface properties can be altered by simple potential application. Questions regarding how the double

layer is affecting the surface properties can then be addressed
by performing parallel experiments in UHV since the SH technique is easily adaptable to both environments. This combination
of experiments is necessary in order to make the direct link between the SH measurements and geometrical structure. Wavelength-dependent studies in both environments are another direction for exploration which have the potential to be extremely
valuable in understanding surface electronic properties.

The three metals examined most extensively to date in the
electrochemical cell are the noble metals: silver, copper, and gold.
All three metals display a rotational anisotropy which can be fit
with the theories discussed above. Furthermore, for Ag(111)
and Cu(111), the anisotropies obtained in solution are qualitatively similar to those obtained for highly ordered crystals under
UHV conditions. The effect of various electrochemical processes
on the anisotropic response will be described in more detail below.

a. Ag(111), Ag(110), and Ag(100)
The rotational anisotropy in the SH intensity from a silver single
crystal surface was first reported by Richmond and co-workers
in both the electrochemical cell [68] and later in UHV [162].
The results for the (p- and s-polarized) SH response from a p-
polarized 1064 nm pump beam incident at an angle ψ of 10° upon
the Ag(111) electrode surface biased in 0.25 $NaClO_4$ near the
PZC is shown in Fig. 16 (a) and (b). The threefold symmetry
of the SH rotational plots reflects the symmetry of the bulk
crystal structure reduced at the boundary by the removal of
the inversion symmetry (C_{3v}). Fits to the data following the
theoretically predicted form (Eqs. (13) and (15)) appear as
solid lines in the figure and the ratio of the constants a and
$c^{(3)}$ is given in the figure caption. The phase of the Fresnel
coefficients accounts for a $\pi/2$ phase shift between the a and
$c^{(3)}$ terms. The p-polarized results show a maximum when the

electric field vector of the incident light, $\bar{E}(\omega)$, is parallel to the
[$2\bar{1}\bar{1}$] direction and a minimum when $\bar{E}(\omega)$ is parallel to the [$11\bar{2}$]
direction. The authors attribute the failure of the s-polarized
SH to vanish completely when ϕ is an integral multiple of $\pi/3$ to
several possible factors including imperfect alignment of the crystal face perpendicular to the plane of incidence, twinning in the
crystal, surface roughness, or the lack of pure polarization in
the input and output beams. Similar results for the Ag(111)
surface have recently been reported in UHV [62,162], although
for the s output data, the minima in the scans approach zero.
Since these Ag single crystal experiments are the first detailed
studies of the SH response from any metallic surface in UHV or
the electrochemical environment, the observations and conclusions
drawn from these studies are given below.

In experiments where the surface is intentionally roughened,
the first effect is the observation of an increase in an isotropic
background for both polarizations after several monolayers of
ORC [54]. With increased roughening, the anisotopic response
becomes isotropic as the surface becomes amorphous in nature.
The p-polarized output was found to be more sensitive to these
roughening effects as would be expected from the fact that the
surface tensor elements responsible for the anisotropy in this
case (Table 1) are ones which contain a component perpendicular
to the surface, χ_{zzz}, χ_{zii}, and χ_{izi}. For the s-polarized output,
the surface tensor elements lie within the plane of the crystal
surface (χ_{xxx}).

The Ag(110) native surface has also been examined at 1064
nm with the results for the p- and s-polarized SH response
shown in Fig. 17a and b, respectively, using a p-polarized pump
beam incident at $\phi = 31°$ [62]. The data in Fig. 17a can be well
fit to the predicted functional form (Eq. (17)) with a proper
selection of the constants a, $c^{(2)}$, $c^{(4)}$, $b^{(2)}$, and $b^{(4)}$. No
anisotropy was observable from the s-polarized output shown in

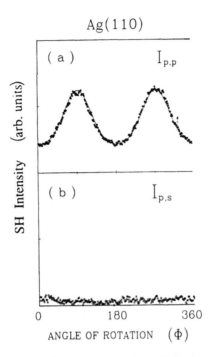

FIG. 17. Second harmonic intensity at 532 nm as a function of angle of rotation for Ag(110) immersed in 0.25 M Na_2SO_4, pH = 5.8 at −0.2 V using a p-polarized pump beam, $\psi = 31°$. (a) P-polarized SH intensity. The theoretical curve (solid line) is generated from Eq. (18) in the text by setting a = 3.1, $c^{(2)}$ = −0.83, and $c^{(4)}$ = 0.08. (b) S-polarized SH intensity. (From Ref. 62.)

Fig. 17b although it was noted that the signal levels from this optical geometry and crystal face were very low.

An important issue in all of these studies is the relative contribution of the bulk anisotropic response to the SH surface response. The SH bulk contribution arises from a layer whose depth is determined not only by the optical skin depth of the material (≈ 400 nm for 532 nm light impinging on a silver surface) but also constrained by the fact that the fields at ω and 2ω

stay coherent for a finite depth [121]. While it is not possible
to separate the bulk contribution from the surface response [122,
123] of a cubic medium, several experimental configurations can
aid in estimating the relative magnitude of these surface and bulk
terms. For example, the bulk ζ term can be separately measured
as the s-polarized SH signal from the (100) face (p- or s-polarized excitation) or the s-polarized signal from the (110) face
(s-polarized excitation). Koos, Shannon, and Richmond [57,62]
have made these three measurements using a 1064 nm fundamental beam and found no anisotropy in the SH response above
the background noise in all cases. Any anisotropy which might
be observed here would have to originate from the bulk of the
crystal. Similar results were also obtained for Ag(100) using a
532 nm excitation beam [57]. Other support for the conclusion
that the bulk contribution is minimal comes from the fits obtained
in Fig. 19 for Ag(110) discussed above. The relative magnitude
of the $c^{(4)}$ term obtained from the fit is small relative to the
$c^{(2)}$ term, which contains surface terms (Table 1).

The observation of rotational anisotropy from silver at these
wavelengths is somewhat surprising since silver bulk interband
transitions are not expected to contribute appreciably to the
nonlinear response, given the dielectric constants measured at
ω and 2ω ($-67.03 + i2.44$ and $-11.9 + i0.33$, respectively). Experimental results clearly indicate that the anisotropic susceptibilities of Ag(111) and Ag(110) do not vanish at these wavelengths. The observation of rotational anisotropy in the second
harmonic response from silver single crystals at these frequencies indicate a deviation from Drude-like behavior of the conduction band electrons. It was speculated in the early work [68]
that surface electronic transitions created by the presence of
water dipoles next to the surface may be playing a role in the
observed anisotropy as has been suggested in linear reflectance
studies [139,163]. The similar anisotropy in the SH response

which has recently been observed for Ag(111) in UHV [162] discounts the possibility that any surface electronic states created by the presence of the aqueous double layer are solely responsible. In addition, rotational anisotropy has been observed for Al in UHV at 1064 nm. This metal is even more free-electron-like in nature at this wavelength [47]. Resonances with low-lying surface states were invoked to explain the observed anisotropy.

The effect of surface electronic resonances have been probed explicitly in the more recent wavelength-dependent studies of Georgiadis, Neff, and Richmond [101]. In these experiments the SH response from Ag(111) and Ag(110) surfaces in an 0.25 M Na_2SO_4 solution at 532 nm input wavelength was examined and compared with the longer wavelength studies and found to be significantly different. Figure 18 displays the rotational anisotropy from p-polarized 532 nm incident light and p- (Fig. 18a) and s- (Fig. 18b) polarized output at 266 nm. Similar data is shown in Fig. 18c and d for the (110) face. The rotational anisotropy of the p-polarized output data for these two crystals is clearly different at 532 nm (Fig. 18a and c) than that obtained for Ag(111) and Ag(110) shown in Figs. 16a and 17a, respectively. These differences are also reflected in the relative magnitudes of the coefficients derived from application of the theoretical fits to the experimental data. For Ag(111) the fits show a shift in phase between a and $c^{(3)}$ as the input wavelength is altered. A similar effect is seen for the p output polarization from the (110) surface. The bulk is not expected to make a significant contribution at this shorter wavelength as noted above.

The authors attribute these changes to an increased coupling to interband transitions in the silver surface which are more important at shorter wavelengths where the d bands of silver can be accessed. This change in phase would result from changes

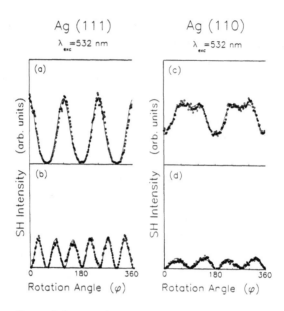

FIG. 18. Second harmonic intensity as a function of angle of
rotation for Ag(111) and Ag(110) immersed in 0.25 M Na$_2$SO$_4$,
pH = 5.0 near the PZC with p-polarized 532 nm illumination.
Angle of incidence (ψ) is 45°. The solid lines are fits to the
data generated from the theoretical expressions given in the
text (Eqs. (13) to (18)). (a) P-polarized SH intensity from
Ag(111), a/c$^{(3)}$ = 1.4. (b) S-polarized SH intensity from
Ag(111). The constant b$^{(3)}$ is taken to be unity. (c) P-
polarized SH intensity from Ag(110) with a/c$^{(2)}$ = −6.8;
a/c$^{(4)}$ = −15. (d) S-polarized SH intensity from Ag(110).
b$^{(2)}$/b$^{(4)}$ = −24. (From Ref. 101b.)

in the Fresnel coefficients of the incoming and outgoing beams
as one approaches an optical resonance with either an interband
transition or surface state. At 1064 nm excitation, both fundamental and second harmonic photon energies are well below the
silver interband transitions (3.8 eV), while at 532 nm excitation,
the SH photon energy is 4.66 eV, well above the interband transition energy. Support for this supposition comes from the
similarity in the observed anisotropy from Ag(111) at the SH

wavelength of 266 nm and the results for Au(111) [55] and
Cu(111) [60,158] at the SH wavelength of 532 nm where optical
resonance can occur because of the lower energy of the d bands
for Cu and Au.

The effect of varying the applied field has also been examined at both wavelengths [62,101]. Stepping the voltage
from −0.6V to −0.2V introduces a large positive electric field
within the first few angstroms of the metal surface. Figure 19a
and b shows the potential dependence in the p- and s-polarized
SH responses at 532 nm from Ag(111) and Ag(110), respectively.
The pump beam was incident upon the sample at an angle ψ =
31° with respect to the surface normal. The (111) and (110)
crystals were held at azimuthal angles of 30° and 45°, respectively, to isolate changes in the isotropic and anisotropic susceptibilities. In the case of Ag(111) (Fig. 19a) the applied
field induces an increase of 130% in both signals. As predicted
in Sec. II.B, the application of this external field does not
change the symmetry of the nonlinear response, only the magnitude of the isotropic and anisotropic responses. A threefold
increase in the isotropic response from Ag(110) is accomplished
by no change in the anisotropic response.

When similar potential dependent experiments are conducted
at 532 nm, the results are dramatically different [101]. Whereas
the potential dependence at 1.064 μm incident excitation is in
agreement with previous work which finds that the SH intensity
is at a minimum at the PZC and rapidly increases with increased
positive charging of the electrode, for 532 nm excitation the behavior is completely reversed. A maximum in the SH intensity
is observed at the PZC followed by a rapid decline as the surface is charged positively (Fig. 10). The experiment shown was
performed at ϕ = 30° relative to the [2$\bar{1}\bar{1}$] direction. The authors attribute this difference to resonant or near resonant
coupling of the radiation field at the shorter wavelength with

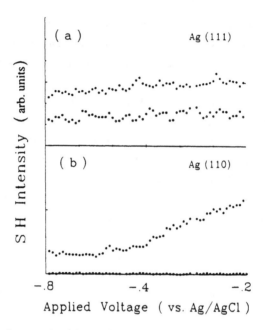

FIG. 19. Isotropic (•) and anisotropic (o) components to the
second harmonic intensity as a function of applied voltage for
silver single crystals immersed in 0.25 M Na_2SO_4, pH = 3.5.
(a) Ag(111), Φ = 30° (b) Ag(110), Φ = 45°. (From Ref. 62.)

surface states on the metal which are altered by variation of the
large dc electric field at the surface. Since the energetic position of surface electronic states is strongly dependent on the
potential at the surface, the authors propose that changes in
the SH response reflect a shift in the energy of surface states
with respect to the Fermi level as the electrode potential is varied. Strongly potential dependent spectral features have been
observed in the optical linear electroreflectance spectroscopy of
various single crystal noble metal electrodes and have been attributed to surface state Stark shifts [138,139]. For the Ag(111)/
electrolyte interface, an energy shift of almost 1 eV/V is reported [138,139]. More importantly, the energy of the state

shifts from about 5eV at −0.20 V bias potential to about 4eV at
−0.80V, moving into a range that is energetically accessible with
532 nm excitation (SH photon energy = 4.6 eV). Thus coupling
with this state is only energetically accessible at more negative
potentials (approaching the PZC) for 532 nm excitation and not
accessible with 1064 nm excitation (SH photon energy = 2.33 eV)
for any potential in the range studied. An analogous explanation
may also account for the behavior of Ag(110), which is reported
to have two surface states [138,139] in the range of our experiments.

As noted above, the surface charge density model [61,103]
has been used to understand the potential dependence of the
SH response at 1064 nm excitation where resonance effects are
not as important. Unfortunately, this simple model cannot take
into account optical resonance effects, which include contributions from surface interband transitions. What is needed is a
theoretical treatment that considers the effect of an applied field
on the detailed bulk and surface electronic band structure such
as has been recently calculated for the metal/vacuum interface
of Ag(001) [136].

A result worth noting regarding these native crystalline
electrode studies is the effect of the presence of an adsorbate
on the surface. Under conditions which are nonresonant with
electronic transitions in the adsorbate, it was found that less
than a monolayer of either an organic impurity or an adsorbing
molecule such as pyrazine would cause a dramatic change in the
rotational anisotropy when 1064 nm is used as a probe wavelength
[164]. Figure 20 demonstrates this effect for Ag(111) in Na_2SO_4
in the absence (Fig. 20a and c) and presence (Fig. 20b and d)
of the impurity. The p-polarized output (Fig. 20a and b) appears to undergo a phase change upon adsorption relative to the
bare polarized surface. The s-polarized output is nearly unaffected (Figs. 20c and d), again displaying the lesser sensitivity

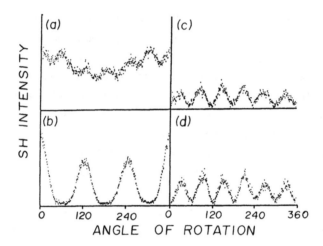

FIG. 20. SH rotational anisotropy from a Ag(111) electrode surface in 0.1 mM $NaClO_4$ using p-polarized 1064 nm input. (a) P-
polarized SH response before and (b) after the potential was
biased near the PZC for 3–4 hours. (c) S-polarized SH response
before and (d) after the potential was biased near the PZC for
3–4 hours. The resultant voltammogram showed approximately
0.5–1 ML of an impurity had adsorbed onto the surface during
this time. (From Ref. 167.)

(relative to the p-polarized output response) of the tensor elements
in this configuration to surface alterations. Due to the unlikely
possibility that the optical measurements are responding to a
well-ordered impurity overlayer, the authors conclude that the
change in the anisotropy is reflecting a change in the electronic
properties of the silver surface upon adsorption.

b. Cu(111) and Cu(100)

The first SH study of Cu(111) was performed by Tom and Aumiller in UHV and in air with a 1064 nm incident beam [158]. A
surface of 3m symmetry was observed as consistent with the
structure of the first 2 layers of the (111) surface. They noted
the importance of resonances between 2ω and the d bands of

copper near 2.1 eV in contributing to the rotational anisotropy
observed for this crystal. The rotational anisotropy from Cu(111)
in simple electrolytes has been examined by Shannon et al. at
1064 nm [60] and both 1064 nm and 532 nm by Robinson et al.
[165]. The results for the 1064 nm incident wavelength is shown
in Fig. 21 for p-polarized fundamental and p- and s-polarized

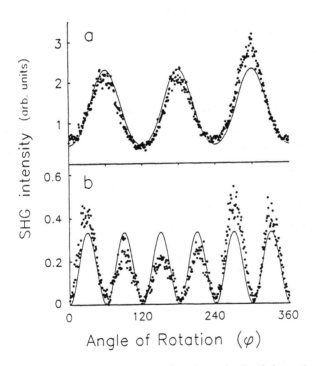

FIG. 21. Second harmonic intensity (p-polarized input) as a
function of angle of rotation for a Cu(111) electrode in 0.25 M
Na_2SO_4 solution at open circuit. The solid lines are fits to the
data using Eqs. (13) and (15); (a) p-polarized SH intensity at
532 nm, $a/c^{(3)} = -2.0$ and (b) s-polarized SH intensity at 532
nm. (From Ref. 60.)

SH light. The fits to the data with respect to Eqs. (13) and (15) are also given. Qualitatively, there is a striking similarity between the UHV results [158] and those obtained in situ [60] suggesting the similarity in surface electronic and structural properties in the two environments, although simultaneous UHV and in situ measurements need to be done to confirm this. The in situ measurements do, however, show a somewhat higher base-line for the p-polarized output which the authors suggest could be due to imperfections on the crystal surface.

As noted above, these results are distinctly different from the silver work for the p-polarized output at this wavelength (compare Fig. 16a and Fig. 21a). The authors note that this difference is not due to geometrical differences in the surface structure of these two different (111) surfaces, but is a result of an optical resonance between 2ω and the d bands of copper as was also suggested in the UHV work. The anisotropic response with 532 nm as a probe wavelength is similar to that obtained for the longer wavelengths [165]. Whereas the relative magnitudes of a, $c^{(3)}$ and $b^{(3)}$ are observed, there is no change in phase between a and $c^{(3)}$ at the shorter wavelengths. Potential dependent effects have also been examined in both of these studies with a variation in the relative magnitudes of the anisotropic and isotropic terms noted. The symmetry of the surface, however, is not observed to change with potential as with Ag(111) and Ag(110).

Cu(100) has been examined by Koos and Richmond [57] to understand the relative bulk and surface contribution from copper crystals at 1064 nm and 532 nm probe wavelengths. As with Ag(100), an experiment conducted with p-polarized input and s-polarized SH output results in a negligible anisotropic response. Since the only anisotropic response for this configuration must come from bulk tensor elements, the bulk contribution was determined to be negligible relative to the surface response.

c. Au(111) and Au(100)

The SH rotational anisotropy was first examined by Koos [55]
for gold single crystals at 1064 nm biased in 0.25 M NaClO4.
The results are similar to those obtained for Cu(111) [60] but
not Ag(111) for the p-polarized output. This similarity with
copper and difference with the silver results was attributed to
an optical resonance between the d bands of Au and the SH
wavelength at 532 nm. The Au(100) studies showed that for
this third noble metal, the surface SH response dominates over
the bulk at this wavelength.

More recently SHG has been used to examine the potential
induced reconstruction of Au(111) and Au(100) in solution [102].
Au(111), which has threefold symmetry, was found to reconstruct
to Au(111) − (1 × 23) as evidenced by a pronounced onefold
symmetry pattern in the SH response shown in Fig. 22. In 0.1
M HClO4 the (1 × 23) structure was stable up to +0.4 V vs. SCE,
while at more positive potentials anion adsorption caused the reconstruction to disappear. Negative charging of the gold surface
fully restored the 1 × 23 structure. The adsorbate-induced
lifting and potential-induced restoration followed first-order kinetics with decay times on the order of 15−20 sec. In a related
study, the anion induced lifting of the Au(100) − (5 × 20) structure and the potential induced reconstruction of Au(100) − (1 ×
1) were studied by SHG also and confirmed the earlier electroreflectance and LEED measurements for this reconstruction [166].

D. SH Response to Metallic Overlayer Deposition

In light of the sensitivity of SHG to double layer perturbations
including variation of electrode potential and adsorption of molecules, it is not surprising that SHG is also very sensitive to
submonolayers of metals deposited or absorbed on the surface of
the substrate metal. The work described in this section has
been aimed at understanding metal overlayer deposition, primarily

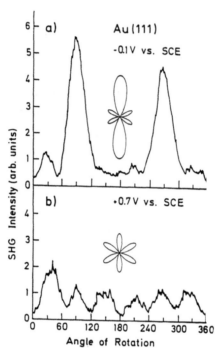

FIG. 22.   Second harmonic intensity at 532 nm as a function of
angle of rotation for Au(111) in 0.01 M H$_2$SO$_4$ at (a) -0.1 V and
(b) $+0.7$ V vs. SCE.   S-polarized excitation and s-polarized SHG
detection in both cases.   For comparison purposes the calculated
polar plots of the rotational anisotropies for s/s-polarization of
a threefold symmetry surface with (a) and without (b) a onefold
symmetry superimposed are also shown.   (From Ref. 102.)

underpotential deposition of various metals on noble metal substrates.   The first studies described will focus on both static
and time-resolved measurements on polycrystalline substrates.
This will be followed by deposition on single crystal substrates
where the SH rotational anisotropy during the deposition will be
examined.

     Since many of the studies to be discussed use a Langmuir
model to describe the adsorption, a few comments about factors

important in using this model to understand the SH response are appropriate. The Langmuir model is applicable under conditions where all of the adsorption sites on the surface are equivalent and noninteracting. Under these conditions, one expects a linear relationship between the change in the nonlinear susceptibility $\chi^{(2)}$ and the surface coverage Θ [60]:

$$\chi^{(2)} = A + B\ (\Theta) \tag{21}$$

where A and B represent the substrate and adsorbate contributions to the nonlinear susceptibility, respectively. Θ is the surface coverage of adatoms with respect to the silver substrate, normalized with respect to the saturation (monolayer) coverage. By including a factor of $e^{i\phi}$, which accounts for the phase of $E(2\omega)$ from the absorbate relative to the field due to the substrate, an expression for the observed intensities as a function of surface coverage can be written as:

$$I(2\omega) = |\ 1 + (2B/A\ (\cos\ \phi_1)\ \Theta + [(B/A)(\Theta)]^2\ | \tag{22}$$

where B/A and ϕ_1 are adjustable parameters.

When the adsorption site changes or interactions between the adatoms change the optical response of the interface, an additional contribution to $\chi^{(2)}$ must be included such that

$$\chi^{(2)} = A + B\Theta_1 + C(\Theta_2 - \Theta_1) \tag{23}$$

The observed intensity can then be written as a quadratic function of $(C/A)(\Theta_2 - \Theta_1)$:

$$\begin{aligned} I(2\omega) = |\ 1 &+ (2\Theta_1 B/A)(\cos\ \phi_1) + (\Theta_1 B/A)^2 \\ &+ 2(C/A)(\Theta_2 - \Theta_1)[\cos\ \phi_2 + (\Theta_1 B/A)\cos(\phi_1 - \phi_2)] \\ &+ [(C/A)(\Theta_2 - \Theta_1)]^2\ | \end{aligned} \tag{24}$$

Again, the data can be fit by using the ratio of the contributions from the substrate and the adatoms, C/A, as a fitting parameter together with an additional phase factor ϕ_2. Subsequent deposition can be treated in an analogous manner. In this method the sign of the phase angle is undetermined. Another assumption made is that the perturbation of the substrate contribution to $\chi^{(2)}$ by the adsorbate is negligible. The validity of this assumption is questionable in cases where at higher coverages, changes in the refractive index of the overlayer will affect substrate contributions to $\chi^{(2)}$ which depend upon the field gradients. This may lead to errors in the calculated values of B/A, C/A, and the associated phase angles. The magnitude of this substrate/adatom interaction may be estimated by combining an actual measurement of the phase of the SH field. Any arbitrariness in assigning the points of abrupt change in θ vs. I^{SH} where steps are introduced in the adsorption isotherms can also be resolved by additional phase measurements.

1. Thin Film Growth on Polycrystalline Substrates

Underpotential deposition (UPD) is the stepwise deposition of a foreign metal on an electrode surface at potentials positive of the Nernstian potential for bulk deposition. Deposition of a monolayer or submonolayer of these foreign metals is intrinsically controlled by the change in the work function of the surface as the overlayer grows in a stepwise manner. Classical electrochemical techniques have been extensively employed to study these systems [167]. The charge passed in each step has been associated with the deposition of a fractional monolayer or with phase transitions in the overlayer structure [168,169]. For some metal/metal systems, a second monolayer is deposited prior to bulk deposition.

Several SH studies of UPD on polycrystalline electrodes have been reported and demonstrate the sensitivity of SHG to metal-adatom interactions. The potential and time dependencies of Pb deposition on Ag thin film electrodes were first studied by Corn and co-workers [61,103]. They found that one monolayer of Pb deposited on a Ag film from a $PbClO_4$ solution decreased the SH intensity compared to that from the same electrode in $NaClO_4$ at the identical potential. The SH signal declined progressively with negative potential prior to full monolayer coverage. The authors noted that the Pb monolayer changes the surface charge density and also affects the surface optical constants. The time-resolved deposition of Pb was also monitored by SHG following a fast potential step. Figure 23 shows the results using a 1064 nm probe beam from a Nd:YAG laser firing at a set delay time after the potential step. The authors observed three distinct processes in the decay of the SH response during deposition. The first portion of the signal decay was assigned to the drop

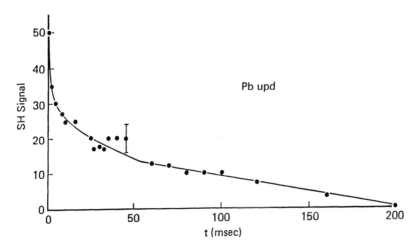

FIG. 23.   Transient SH response from a silver thin film electrode during Pb monolayer formation.   Potential step from −250 mV to −450 mV in 0.1 M sodium acetate and 5 mM lead acetate.   The curve illustrates the three mechanisms discussed in the text. (From Ref. 103.)

in the dc field, the second to diffusion of ions to the electrode, and the longer decay to kinetic controlled formation of the metal monolayer.

Furtak, Miragliotta, and Korenowski [58] were the first to examine to thallium deposition on a polycrystalline surface during a potential sweep. This was followed by a more detailed time-resolved measurement of the thallium deposition by Richmond and Robinson [59] using a mode-locked Nd:YAG laser operating in a "continuous" mode after a fast potential step. Figure 24 (upper trace) shows the CV recorded for polycrystalline Ag with the underpotential region anodic of bulk thallium deposition (−0.78 V) [59]. The first three cathodic (A_1 through A_3) peaks are associated with formation of the first thallium monolayer and correspond to respective coverages of approximately $\Theta = 0.33$, and $\Theta = 0.68$, and $\Theta = 1$. The peaks A_1 through A_3 are believed to arise from deposition at energetically different sites on the electrode surface. This structure is not observed on electrodes which are prepared by mechanical polishing [58], yielding instead a single broad peak in the UPD region. The peak B indicates the formation of a second monolayer prior to bulk deposition.

The underpotential deposition of thallium significantly alters the nonlinear susceptibility of the interface. Figure 24 (lower trace) displays the potential dependent SH intensity recorded for a scan from −0.10V to −0.76 V where the second monolayer is formed [59]. The SH intensity initially decreases due to the previously noted DC electric field effect. The slope increases abruptly at about −0.43 V to −0.45 V, in the region where thallium deposition begins (peak A_1 in Fig. 24.) Deposition at site A_1 is complete at −0.48 V. For smooth electrodes in a simple electrolyte, the minimum in the SH intensity is associated with the zero charge condition (−0.92 V for this system) at 45° incidence. Thus the thallium deposit modifies the nonlinear

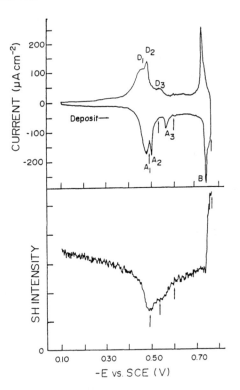

FIG. 24. Cyclic voltammogram and second harmonic response at 532 nm for a polycrystalline Ag electrode immersed in 0.25 M NaClO$_4$ and 1 mM Tl$_2$SO$_4$, pH 3.5, sweep rate 10 mV/sec. The arrows in the SH figure indicate the final potential, E$_f$, of the potential step from −0.10 V used in the time resolved experiments. (From Ref. 59.)

optical properties of the interface, leading to a minimum at $\Theta \approx$ 1/3 ml (−0.48 V). As deposition continues up to $\Theta = 1$ ML, the SH intensity rises with a slight change in slope between the deposition at site A$_2$ ($\Theta \approx 2/3$ ML) and at site A$_3$ ($\Theta = 1$ ML). 150 to 200 percent enhancement in the SH response is observed upon deposition of the second monolayer. The onset of the enhancement corresponds to the beginning of peak B in the CV of Fig.

24. The voltammogram reported by Furtak et al. [58] exhibited somewhat broader peaks but also gave the enhancement at 2 ML. Both studies attributed the enhancement to a resonance between the incoming wavelength and an electronic transition in bulk thallium at 2 ML.

Robinson and Richmond [59] monitored both the time-dependent SH intensity and total cell current in response to a fast potential step between −0.10 V and potential just cathodic of each of the four deposition peaks in the CV (Fig. 24.) The time-resolved SH data provide insight into the deposition behavior that is not readily discerned by standard current transient analysis. Figure 25 displays the time-dependent SH signals observed for thallium deposition at sites A_1 through A_3 and site B. The first transient (Fig. 25a) monitors the time-dependent nonlinear optical properties of filling site A_1 to coverage $\theta \approx 1/3$ ML. Two components are readily discerned in the data: a rapid fall prior to about 10 msec and a much slower decay which has almost reached a minimum by 180 msec. The rapid fall in Fig. 25a (labeled DC) monitors the effects of the decreasing DC electric field and concurrent perchlorate anion desorption. The slower decay of the transient in Fig. 25a from 10 to 180 msec monitors the UPD process. Thus the charging (nonfaradaic) and UPD (faradaic) reactions occur on sufficiently different time scales that the kinetics of the two processes can be determined from the data.

The SH transients in Fig. 25b through d correspond to increasing Tl coverage. The data indicates that deposition proceeds in a successive manner. For example, for coverage $\theta \approx 2/3$ ML, the second transient shows a rapid fall followed by a decay to a minimum in about 150–180 msec, as observed for deposition at site A_1, before the SH signal increases as site A_2 is covered. A third process is apparent at longer times in Fig. 25c, which follows the formation of the complete monolayer at

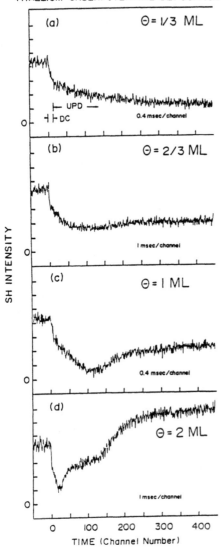

THALLIUM UNDERPOTENTIAL DEPOSITION

(a) $\Theta = 1/3$ ML

├─ UPD ─┤
┤├ DC   0.4 msec/channel

(b) $\Theta = 2/3$ ML

I msec/channel

(c) $\Theta = 1$ ML

0.4 msec/channel

(d)

$\Theta = 2$ ML

I msec/channel

SH INTENSITY

0   100   200   300   400
TIME (Channel Number)

FIG. 25. Time-dependent second harmonic response for a poly-crystalline Ag electrode in $Tl_2SO_4/NaClO_4$ solution for potential steps from −0.10 V to E_f. (a) Submonolayer deposition, E_f = −0.48 V, 250 msec deposit, 150 msec strip. (b) Submonolayer deposition, E_f = 0.53 V, 500 msec deposit, 200 msec strip. (c) One monolayer deposition, E_f = −0.60 V, 200 msec deposit, 150 msec strip. (d) Two monolayer deposition, E_f = −0.76 V, >450 msec deposit, 150 msec strip. (From Ref. 59.)

site A_3. The stepwise formation of the deposits is even more clear for the case of 2 ML deposition (Fig. 25d), which shows that the second overlayer begins immediately after completion of the first layer at about 140 msec. The onset of each sequential deposition step, as indicated by the SH response, moves to shorter times as the magnitude of the potential step increases. This result suggests that the first site is filled faster in Fig. 25b through d, indicating that an activation barrier to reaction may be more easily overcome by the increasing overpotential.

Robinson and Richmond [59] found that the SH transients in Fig. 25 do not follow Langmuir kinetics and are not controlled by diffusion of thallium to the surface. The results of a phenomenological fitting analysis are shown for the submonolayer transients (Fig. 26). The SH data obey exponential rate laws where the number of exponentials equal the number of sites filled in agreement with the current transients, and verifying the sequential, rather than competitive, deposition. The exponential form of i(t) indicates that adsorption is the deposition mechanism. The SH data decay faster than the current transients for the first two sites. This result indicates that surface migration of thallium adatoms does not contribute to the kinetics, as this scenario would require that the SH transients decay more slowly than the current transients. Conversely, the SH data suggests filling the third site is slower than the current transient suggests.

Campbell and Corn have also examined the deposition of Li on polycrystalline Ag [79]. In order to attain the negative voltages necessary for the alkali deposition, an acetonitrile solution containing 0.1 M $LiClO_4$ was employed. Although the UPD of lithium on the silver substrate does occur, due to the reactivity of the metallic lithium with impurities in solution, the adsorbed layer formed on the negative potential sweep is not as stable as other UPD monolayers. During the deposition, the second

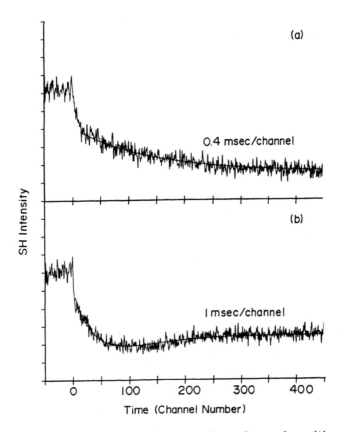

FIG. 26. Phenomenological fits to submonolayer deposition SH transients. (a) Single exponential fit to deposition at first site. $\tau_1 = 55 \pm 2$ msec. (b) Difference of two exponential fits to deposition at first and second sites. $\tau_1 = 55 \pm 8$ msec; $\tau_2 = 53 \pm 8$ msec. (From Ref. 59.)

harmonic signal at 532 nm is found to increase. The authors attribute this change in signal to an increase in free electron density on the surface through the delocalization of the lithium 2s electron.

2. SH Rotational Anisotropy Studies of Deposition
on Crystalline Substrates

The underpotential deposition of metal overlayers on the single crystal substrates of these three noble metals have been the subject of several studies. The goal of all of these studies has been to attempt to correlate the rotational anisotropy observed with the electronic and structural properties of the substrate and overlayer in additional to obtaining quantitative information about changes in the tensor elements associated with the observed SH response.

The most extensive work to date has been for thallium deposition on silver single crystals. Koos, Shannon, and Richmond [62] examined Tl deposition on Ag(111) and Ag(110) using a Q-switched Nd:YAG laser followed by Miragliotta and Furtak [52] in which a cw Nd:YAG laser was employed. The behavior of thallium deposition on Ag(111) is quite complex as indicated by the structure observed in the voltammetry (Fig. 27) [62]. The first three deposition peaks in the CV ($A_1 - A_3$) are assumed by previous studies to correspond to the sequential formation of two intertwined $(\sqrt{3} \times \sqrt{3})R30°$ overlayers which are subsequently filled in to form a complete monolayer before bulk deposition takes place [152,170,171]. A second monolayer (A_4) is then formed before bulk deposition occurs (−0.8 V).

The rotational anisotropy in the p-polarized SH response obtained at various stages of deposition is shown in Fig. 28a to c and can be compared with the native silver surface shown in Fig. 16a. Deposition of thallium up to the first monolayer (Fig. 28a) modifies the rotational scan in a dramatic manner. The 3m symmetry expected from the (111) surface is retained and a

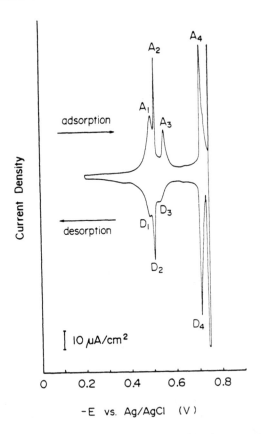

FIG. 27. Cyclic voltammogram for Ag(111) immersed in 0.25 M
Na_2SO_4 and 5 mM Tl_2SO_4, pH = 3.2, sweep rate = 10 mV/sec.
(From Ref. 3.)

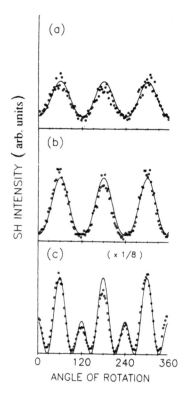

FIG. 28. P-polarized second harmonic intensity as a function of
angle of rotation for a Ag(111) electrode with various thallium
coverages. (a) $0 < \Theta < 1$ monolayer, (b) $\Theta = 1$ monolayer, and
(c) $\Theta = 2$ monolayers. (From Ref. 3.)

phase shift is introduced between a and $c^{(3)}$. When a second
monolayer of thallium is adsorbed, the relative magnitudes of the
susceptibility elements are noticeably altered as the ratio of the
isotropic and anisotropic terms changes from $| a/c^{(3)} | > 1$ to
$| a/c^{(3)} | < 1$. Furthermore, the signal levels are enhanced by
approximately 9 times those observed in the nonspecifically adsorbing electrolyte at the same potential.

A more insightful analysis of these alterations in the surface electronic properties can be obtained by fixing the electrode surface and varying the potential throughout the deposition region [62]. Separate determination of changes in the isotropic and anisotropic components of the surface susceptibility tensor can be obtained by monitoring the SH intensity at a fixed angle of $\phi = 30°$, where $I^{SH}_{\parallel,\parallel} \propto |a|^2$ and $I^{SH}_{\parallel,\perp} \propto |b^{(3)}|^2$. Figure 29 shows the results derived from such experiments where the SH response from the isotropic (Fig. 29a) and anisotropic

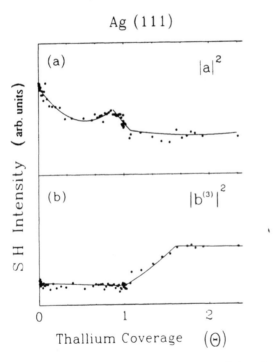

$$\text{Ag}\ (1\,1\,1)$$

FIG. 29.   Second harmonic response at $\phi = 30°$ for Ag(111) immersed in 5 mM Tl_2SO_4 as a function of thallium coverage (Θ). Incident wavelength = 1064 nm.   (a) P-polarized signal where $I^{SH} \propto |a|^2$.   (b) S-polarized signal where $I^{SH} \propto |b|^2$.   The solid lines are fits to the data according to Eq. (24) in the text. (From Ref. 62.)

components (Fig. 29b) are plotted as a function of thallium coverage (θ). Fits to the data were obtained using Eqs. (21) to (24).
Changes in the magnitude of the isotropic susceptibility a, shown
in Fig. 29a, reflect the complicated nature of deposition process.
The results can be modeled by assuming three distinct stages to the
deposition process. From θ = 0 to θ = 0.87 (−0.35 to −0.54 V in the
CV), the isotropic signal is modified by the depositing thallium
atoms continuously with B/A = 1.12 and ϕ_1 = 128.8°. During
the deposition peak labeled A_3, an abrupt change is observed
where further contribution from the adatoms can be modeled by
setting C/A = −0.80 and ϕ_2 = 63.5°. Once the first layer is
complete, θ = 1.1, as determined by the coulometry, the isotropic terms remain rather constant through the formation of the
second adlayer. The magnitude of the anisotropic susceptibility
$b^{(3)}$ (Fig. 29b) is relatively unperturbed by the formation of
the first monolayer ($A_1 - A_3$). This is followed by significant
enhancement in the magnitude of $b^{(3)}$ once the second monolayer
begins to form, which continues until the second monolayer is
2/3 complete.

Studies of thallium deposition on Ag(111) have also been
performed by Miragliotta and Furtak [52] in which similar observations have been made although the sensitivity to changes
in the submonolayer coverages is lower. The unique aspect of
this later work is the use of an inexpensive cw Nd:YAG laser.
They, however, did not observe an anisotropy from the adsorbate
free Ag(111) surface under the normal incidence optical geometry
that they performed their experiments.

Koos et al. [62] also examined the electrochemical deposition
of thallium on Ag(110) and found results similar to those obtained
from the (111) face. The voltammogram shows well-defined structure in the formation of the first monolayer and further deposition occurs before formation of the bulk deposit. The SH response depicts a well-ordered overlayer growth for up to 2
monolayers which displays 2m symmetry.

The studies from both groups suggest that the enhanced anisotropic signals seen for Ag(111) [52,62] and Ag(110) [62] at coverages greater than $\Theta = 1$ ML result from the optical properties of the thallium thin film. An enhancement of the anisotropic nonlinear response would be expected if photons of either energy can couple to an eigenstate of the overlayer. Bulk interband transitions are predicted by theory and observed experimentally for both lead and thallium in this wavelength region [172]. The enhancement in the anisotropic component at 2 ML observed here is most likely due to such a resonance between ω or 2ω and bulk interband transitions of thallium which arise as the overlayer begins to develop a two-dimensional structure.

A somewhat simpler system than Ag(111)/Tl^{2+} is the Ag(111)/Pb^{2+} UPD [62]. The voltammogram shows a sharp peak prior to bulk deposition at -0.52 V which yields a value of 337 (± 20) $\mu C/cm^2$ for the charge passed or approximately 1 ML. The SH rotational anisotropy obtained upon 1 ML deposition is shown in Fig. 30. A significant change in the nonlinear response of the interface is observed in the p-polarized output (Fig. 30a) upon adsorption of lead with the data indicating that the interfacial region retains 3m symmetry. A fit to the data is obtained by setting $a/c^{(3)} = -2.7$ and $b^{(3)} = 0.65$. A clear change in the relative phase of a and $c^{(3)}$ is indicated. The s-polarized data (Fig. 30b) is well fit by the theory where only the anisotropic terms contribute. A 37% decrease in the value of a and a 43% decrease in the value of $c^{(3)}$ compared to those measured for the adsorbate free surface are observed.

Thallium deposition on Cu(111) [60] provides an interesting comparison with the Ag(111) data due to the larger mismatch in the lattice constants between Cu and Tl relative to the more closely matched Tl and Ag. In this study by Shannon, Koos, and Richmond [60] using 1064 nm as the fundamental, an abrupt change in the adsorption isotherm for the isotropic contribution

Ag(1 1 1)/Pb

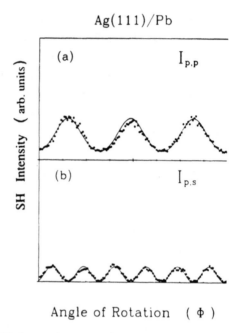

Angle of Rotation   (Φ)

FIG. 30.   SH intensity as a function of angle of rotation for Ag(111) in NaClO$_4$ containing Pb^{2+} at a lead coverage of one monolayer, E = −0.46V.   Solid lines are fits to the data using Eqs. (13) and (15) found in the text.   (a) P-polarized SH, a/c$^{(3)}$ = 2.7.   (b) S-polarized SH, b$^{(3)}$ = 0.65.   (From Ref. 62.)

to $\chi^{(2)}$ was noted at a coverage which corresponded to the formation of an epitaxial overlayer, θ = 1/3.   A comparison between the two systems suggests that at higher coverages on the Cu(111) substrate the lattice mismatch results in the formation of the first monolayer as an either buckled or incommensurate overlayer. This leads to abrupt changes in the isotropic components of the nonlinear optical response of the interface.   The anisotropic response during the deposition does not provide conclusive evidence for the formation of bulk thallium at near 2 ML as was seen for Ag(111) and is under further investigation.   A time-

resolved measurement of deposition kinetics in this system has
been done which isolates the time-dependent changes in the isotropic and anisotropic contribution to the SH response during
thin film formation [173].

Thallium deposition on Au(111) [55] has been examined by Koos
in the same manner as the Ag(111) and Cu(111) at 1064 nm.
The lattice constant for this metal is similar to that of Ag.   It
was found in this study that upon one monolayer of deposition,
the rotational anisotropy from the previously clean Au(111) surface
disappears and an isotropic response results. With two monolayers, the anisotropy returns, indicating an epitaxial layer of
Tl on the Au(111) surface.

The electrodeposition of Cd and Zn on Ag(111) has been
reported by Miragliotta and Furtak [52], again using a contin-
uous-wave Nd:YAG laser operating at 1064 nm and normal incidence. For the Cd system, the UPD of up to one monolayer
proceeds by a three-step process as indicated in the voltammogram. In the SH experiments, the intensity was found to increase when 1 ML was formed, reaching a maximum at 1.4 ML.
They then observed a decrease at 2.0 ML, with the SHG intensity remaining unchanged through coverages up to 3.0 ML. The
rotational anisotropy was then recorded at the 1.4 and 2.0 ML
coverages. At a deposition of 1.4 ML, the SH response gives
C_{3v}m symmetry. No rotational anisotropy was observable on
the native surfaces. The model which is consistent with their
observations is shown in Fig. 31. They propose that each 1/3
ML adopts the ($\sqrt{3} \times \sqrt{3}$) R 30° structure. Due to differences
in Cd coordination, each step is presumed to involve successively
lower binding energies and, as a result, resides at a slightly
different height above the Ag(111) surface. This would then
permit epitaxy and keep the three parts of the full monolayer
. equivalent and the resulting buckled structure would have the
necessary C_{3v} symmetry. They attribute the large increase in

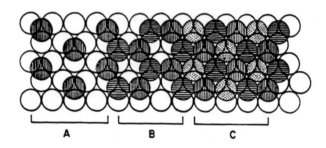

FIG. 31. Model of Cd deposition in the underpotential region.
Coverage corresponds to (A) $\Theta = 1.3$ ML; (B) $\Theta = 2/3$ ML; (C)
$\Theta = 1$ ML. (From Ref. 53.)

intensity observed at 1 ML to a resonance between the incident
photons at 1.17 eV and pseudobasal plane Cd through parallel
transitions. Reflectance studies of bulk single crystal Cd have
shown interband transitions occur at 1 eV. With two full monolayers, the overlayer appears to have C_1 symmetry. They attribute this to a close-packed hexagonal two-layer Cd lattice
that is incommensurate with Ag, providing that the Ag continues
to influence the electronic structure of at least the first monolayer.

In the Zn studies by this group [52], a similar intensity
increase was found at the potential corresponding to the onset
of the second ML of Zn. For a Zn coverage between 1.6 ML and
3.1 ML, the SHG remained constant. However, at more negative
potentials there was a noted increase in SHG, which was coincident with the onset of the hydrogen generation reaction. The
energy of interband transitions for single crystal Zn appears
between 1.2 eV and 1.5 eV. The authors therefore conclude
that the lack of observable SHG before the first monolayer is complete suggests that hybridization among Zn adatoms is needed to develop the proper energies to support electronic resonances. The
rotational anisotropy patterns suggest that Zn is deposited in an
epitaxial 8.6% expanded hexagonal monolayer for Zn, followed by

a return to the Zn lattice constant at higher coverages. They compared this with the Tl work, which they proposed grows with a coincidence lattice that is (6 × 6) − 25 Tl for all coverages studied up to 5 ML.

V. CONCLUDING REMARKS

This review has been written to provide a summary of the current state of the field of surface second harmonic generation as applied to electrochemical systems. Examples of theoretical and experimental studies which probe fundamental issues regarding the source of the nonlinear polarizability at the solid/liquid interface have been presented. Complementary studies employing SHG as a tool for measuring the structural and thermodynamic properties of components in the electrochemical double layer, as well as dynamic information, have also been detailed. Even with the significant advances made through these efforts over the past 10 years, the field is still in a very early stage of development both in UHV and in solution. Essential to its progress is the need for a clearer understanding and theoretical description of the source of the nonlinear polarizability as it relates to microscopic properties of the interfacial components. Other experiments are needed to compare the nonlinear optical response from the electrode surface in both UHV and in situ environments in order to obtain a clearer understanding of the correlation between the rotational anisotropy and structural aspects of the electrode surface. Further wavelength-dependent studies would also be valuable with the eventual goal of being able to use SHG to characterize surface states in situ as has been done by other methods in UHV. Nearly all of the experiments described in the electrochemical cell have been on metallic systems with the characterization of semiconductor and modified electrodes yet to be examined. Progress in all of these areas of study would not only enhance the use of SHG as a general analytical tool for many technologically important areas, but would also contribute to the advancement of other new nonlinear optical techniques which are interface specific such as sum-frequency generation.

ACKNOWLEDGMENTS

The author expresses her appreciation to D. A. Koos, R. Georgiadis, R. Bradley, and J. M. Robinson for their help with this manuscript. Financial support from the Army Research Office, the Office of Naval Research, and the Petroleum Research Fund of the American Chemical Society (PRF-20274AC5) is gratefully acknowledged. The author also thanks the Henry and Camille Dreyfus Teacher-Scholar Award and the NSF Presidential Young Investigator Award for additional support of this work.

REFERENCES

1. P. A. Franken, A. E. Hill, C. W. Peters, and G. Weinreich, Phys. Rev. Lett. $\underline{7}$:118 (1961).

2. Y. R. Shen, The Principles of Nonlinear Optics, Wiley, New York, 1984.

3. G. L. Richmond, J. M. Robinson, and V. L. Shannon, Prog. in Surf. Sci. $\underline{28}$:1 (1988).

4. Y. R. Shen, Nature $\underline{337}$:519 (1989).

5. N. Bloembergen and P. S. Pershan, Phys. Rev. $\underline{128}$:606 (1962).

6. J. Ducuing and N. Bloembergen, Phys. Rev. Letters $\underline{10}$: 474 (1963).

7. F. Brown, R. E. Parks, and A. M. Sleeper, Phys. Rev. Lett. $\underline{14}$:1029 (1965).

8. S. S. Jha, Phys. Rev. $\underline{140}$:A2020 (1965); S. S. Jha, Phys. Rev. Lett. $\underline{15}$:412 (1965).

9. N. Bloembergen and Y. R. Shen, Phys. Rev. $\underline{141}$:298 (1966).

10. F. Brown and R. E. Parks, Phys. Rev. Lett. 16:507 (1966).

11. N. Bloembergen, R. K. Chang, and C. H. Lee, Phys. Rev. Lett. 16:986 (1966).

12. S. S. Jha and C. S. Warke, Phys. Rev. 153:751 (1967).

13. C. H. Lee, R. K. Chang, and N. Bloembergen, Phys. Rev. Lett. 18:167 (1967).

14. H. Sonnenberg and H. Heffner, J. Opt. Soc. Am. 58:209 (1968).

15. G. V. Krivoshchekov and V. I. Stroganov, Sov. Phys. - Solid State 9:2856 (1968).

16. G. V. Krivoshchekov and V. I. Stroganov, Sov. Phys. - Solid State 11:89 (1969).

17. C. C. Wang and A. N. Duminski, Phys. Rev. Lett. 20:668 (1968).

18. R. W. Terhune, P. D. Maker, and C. M. Savage, Phys. Rev. Lett. 8:404 (1962).

19. N. Bloembergen, R. K. Chang, S. S. Jha, and C. H. Lee, Phys. Rev. 174:813 (1968).

20. C. C. Wang, Phys. Rev. 178:1457 (1969).

21. F. Brown and M. Matsuoka, Phys. Rev. 185:985 (1969).

22. J. Rudnick and E. A. Stern, Phys. Rev. B 4:4274 (1971).

23. J. R. Bower, Phys. Rev. B 14:2427 (1976).

24. G. V. Krivoshchekov and V. I. Stroganov, Sov. Phys. - Solid State 11:2151 (1970).

25. C. S. Wang, J. M. Chen and J. R. Bower, Opt. Comm. 8:275 (1973).

26. J. M. Chen, J. R. Bower, C. S. Wang, and C. H. Lee, Opt. Comm. 9:132 (1973).

27. H. J. Simon, D. E. Mitchell, and J. G. Watson, Phys. Rev. Lett. 33:1531 (1974).

28. H. J. Simon, R. E. Benner, and J. G. Rako, Opt. Comm. 23:245 (1977).

29. D. L. Mills, Solid State Comm. 24:669 (1977).

30. M. Fukui, V. So, and G. I. Stegeman, Phys. Rev. B 18:2484 (1978).

31. J. E. Sipe, V. C. Y. So, M. Fukui, and G. I. Stegeman, Phys. Rev. B 21:4389 (1980).

32. M. Fukui and G. I. Stegeman, Solid State Comm. 26:239 (1978).

33. Y. J. Chen and E. Burstein, Il Nuovo Cimento 39B:807 (1977).

34. H. J. Simon, D. E. Mitchell, and J. G. Watson, Opt. Comm. 13:294 (1975).

35. M. Fukui, V. C. Y. So, J. E. Sipe, and G. I. Stegeman, J. Phys. Chem. Solids 40:523 (1979).

36. C. K. Chen, A. R. B. deCastro, and Y. R. Shen, Phys. Rev. Lett. 46:145 (1981).

37. R. K. Chang and T. E. Furtak, Surface Enhanced Raman Scattering, Plenum Press, New York, 1982.

38. S. G. Grubb, A. M. DeSantolo, and R. B. Hall, J. Phys. Chem. 92:1419 (1988).

39. H. W. K. Tom, C. M. Mate, X. D. Zhu, J. E. Crowell, T. F. Heinz, G. A. Somorjai, and Y. R. Shen, Phys. Rev. Lett. 52:348 (1984).

40. H. W. K. Tom, C. M. Mate, X. D. Zhu, J. E. Crowell, Y. R. Shen, and G. A. Somorjai, Surf. Sci. 172:466 (1986).

41. D. Heskett, K. J. Song, A. Burns, E. W. Plummer, and H. L. Dai, J. Chem. Phys. 85:7490 (1986).

42. D. Heskett, L. E. Urbach, K. J. Song, E. W. Plummer, and H. L. Dai, Surf. Sci. 197:225 (1988).

43. X. D. Zhu, Y. R. Shen, and R. Carr, Surf. Sci. 163:114 (1985).

44. C. C. Mate, G. A. Somorjai, H. W. K. Tom, X. D. Zhu, and Y. R. Shen, J. Chem. Phys. 88:441 (1988).

45. H. W. K. Tom, X. D. Zhu, Y. R. Shen, and G. A. Somorjai, Surf. Sci. 167:167 (1986).

46. W. Hener, L. Schröder, and H. Zacharias, Chem. Phys. Lett. 135:299 (1987).

47. R. Murphy, M. Yeganch, K. J. Song, and E. W. Plummer, Phys. Rev. Lett. 63:318 (1989).

48. K. J. Song, D. Heskett, H. L. Dai, A. Kiebsch, and E. W. Plummer, Phys. Rev. Lett. 61:1380 (1988).

49. J. M. Hicks, L. E. Urbach, E. W. Plummer, and Hai-Lung Dai, Phys. Rev. Lett. 61:2588 (1988).

50. J. D. Hamilton and R. J. M. Anderson, Chem. Phys. Lett. 151:455 (1988).

51. R. J. M. Anderson and J. C. Hamilton, Phys. Rev. B. 38:8451 (1988).

52. J. Miragliotta and T. E. Furtak, Surf. Interface Anal. 14:53 (1989).

53. J. Miragliotta and T. E. Furtak, Phys. Rev. B. 37:1028 (1988).

54. V. L. Shannon, D. A. Koos, J. M. Robinson, and G. L. Richmond, in Chemically Modified Surface, Vol. 2, Gordon and Breach Science Publishers, New York, 1988, p. 485.

55. D. A. Koos and G. L. Richmond, J. Electrochem. Soc. 136:218C (1989).

56. V. L. Shannon, J. M. Robinson, and G. L. Richmond, Spectroscopy 3:4 (1988).

57. D. A. Koos and G. L. Richmond, Bulk Contribution from the SH Response from Noble Metal Surfaces, Phys. Rev., in press.

58. T. E. Furtak, J. Miragliotta, and G. M. Korenowski, Phys. Rev. B. 35:2569 (1987).

59. J. M. Robinson and G. L. Richmond, Time resolved measurement of the electrodeposition of thallium on silver electrodes by second harmonic generation, Chemical Physics, in press.

60. V. L. Shannon, D. A. Koos, S. A. Kellar, and G. L. Richmond, J. Phys. Chem. 93:6434 (1989).

61. R. M. Corn, M. Romagnoli, M. D. Levenson, and M. R. Philpott, Chem. Phys. Lett. 106:30 (1984).

62. D. A. Koos, V. L. Shannon, and G. L. Richmond, J. Phys. Chem., 94:2091 (1990).

63. T. F. Heinz and M. M. T. Loy, Adv. Laser Sci. 3:452 (1988).

64. T. F. Heinz, M. M. T. Loy, and S. S. Iyer, Mater. Res. Soc. Symp. Proc. 75:697 (1987).

65. G. Berkovic and Y. R. Shen, in Nonlinear Optical and Electroactive Polymers: Proc. of the 193rd Nat. Am. Chem. Soc., Plenum, New York, 1987.

66. V. L. Shannon, D. A. Koos, and G. L. Richmond, J. Phys. Chem. $\underline{91}$:5548 (1987).

67. V. L. Shannon, D. A. Koos, and G. L. Richmond, Appl. Opt. $\underline{26}$:3579 (1987).

68. V. L. Shannon, D. A. Koos, and G. L. Richmond, J. Chem. Phys. $\underline{87}$:1440 (1987).

69. G. L. Richmond, H. M. Rojhantalab, J. M. Robinson, and V. L. Shannon, J. Opt. Soc. Am. B. $\underline{4}$:228 (1987).

70. G. L. Richmond, Chem. Phys. Lett. $\underline{110}$:571 (1984).

71. G. L. Richmond, Langmuir $\underline{2}$:132 (1986).

72. G. L. Richmond, Chem. Phys. Lett. $\underline{106}$:26 (1984).

73. H. M. Rojhantalab and G. L. Richmond, J. Phys. Chem. $\underline{93}$:3269 (1989).

74. H. M. Rojhantalab and G. L. Richmond, in Advances in Laser Science - II, AIP Conference Proceedings Series No. 180, (W. C. Stwalley, M. Lapp, and G. A. Kenney-Wallace, eds.), AIP, New York, 1987.

75. J. M. Robinson, H. M. Rojhantalab, V. L. Shannon, D. A. Koos, and G. L. Richmond, Pure and Appl. Chem. $\underline{59}$:1263 (1987).

76. P. Chu and G. L. Richmond, "Comparative studies of the Ag/aqueous electrolyte interface by photoacoustic and nonlinear optical methods," J. Electroanal. Chem., in press.

77. D. F. Voss, M. Nagumo, L. S. Goldberg, and K. A. Bunding, J. Phys. Chem. $\underline{90}$:1834 (1986).

78. D. J. Campbell and R. M. Corn, J. Phys. Chem. $\underline{91}$:5668 (1987).

79. D. J. Campbell and R. M. Corn, J. Phys. Chem. $\underline{92}$:5796 (1988).

80. D. A. Koos, V. L. Shannon, J. M. Robinson, and G. L. Richmond, in Proceedings of the 173rd Meeting of the Electrochem. Soc., Atlanta, GA, May 1988.

81. B. M. Biwer, M. J. Pellin, M. W. Schauer, and D. M. Gruen, Surf. Sci. $\underline{176}$:377 (1986).

82. V. L. Shannon, D. A. Koos, J. M. Robinson, and G. L. Richmond, Chem. Phys. Lett. $\underline{142}$:323 (1987).

83. C. K. Chen, T. F. Heinz, D. Ricard, and Y. R. Shen, Phys. Rev. Lett. $\underline{46}$:1010 (1981).

84. C. K. Chen, T. F. Heinz, D. Ricard, and Y. R. Shen, Phys. Rev. B. 27:1965 (1983).

85. D. V. Murphy, K. U. von Raben, R. K. Chang, and P. B. Dorain, Chem. Phys. Lett. 85:43 (1982).

86. T. F. Heinz, C. K. Chen, D. Ricard, and Y. R. Shen, Chem. Phys. Lett. 83:180 (1981).

87. G. L. Richmond, Surf. Sci. 147:115 (1984).

88. D. V. Murphy, K. U. von Raben, T. T. Chen, J. F. Owen, R. K. Chang, and B. L. Laube, Surface Science, 124:529 (1983).

89. C. D. Marshall and G. M. Korenowski, J. Chem. Phys. 85:4172 (1986).

90. C. D. Marshall and G. M. Korenowski, J. Phys. Chem. 91:1289 (1987).

91. T. T. Chen, K. U. von Raben, D. V. Murphy, and R. K. Chang, Surf. Sci. 143:369 (1984).

92. G. L. Richmond, Chem. Phys. Lett. 113:359 (1985).

93. G. L. Richmond and P. Chu, Proc. of the 189th Nat. Am. Chem. Soc. Meeting, Miami, FL, May 1985.

94. D. J. Campbell and R. M. Corn, ACS Symp. Ser. 378:294 (1988).

95. J. M. Robinson and G. L. Richmond, Electrochim ACTA, 34:1639 (1989).

96. M. W. Schauer, M. J. Pellin, and D. M. Gruen, Chem. Mater. 1:101 (1989).

97. D. J. Campbell and R. M. Corn, Molecular second harmonic generation studies of methylene blue chemisorbed onto a sulfur modified polycrystalline platinum electrode, J. Phys. Chem., in press.

98. B. M. Biwer, M. J. Pellin, M. W. Schauer, and D. M. Gruen, Electrochemical and second harmonic generation investigation of nickel corrosion in 0.1 M NaOH, submitted.

99. G. L. Richmond, Nonlinear optical studies of electrochemical interfaces, Proceedings of FAACS, Detroit, MI, Sept. 1986.

100. B. M. Biwer, M. J. Pellin, M. W. Schauer, and D. M. Gruen, Langmuir, 4:121 (1988).

101a. R. M. Georgiadis, G. A. Neff, and G. L. Richmond, J. Chem. Phys. 92:4623 (1990).

101b. R. M. Georgiadis, G. A. Neff, and G. L. Richmond, in preparation.

102. A. Friedrich, B. Pettinger, D. M. Kolb, G. Lupke, R. Steinhoff, and G. Marowsky, Chem. Phys. Lett. 163:123 (1989).

103. R. M. Corn, M. Romagnoli, M. D. Levenson, and M. R. Philpott, J. Chem. Phys. 81:4127 (1984).

104. H. L. Hampsch, J. Yang, G. K. Wong, and J. M. Torkelson, Macromolecules 21:528 (1988).

105. J. M. Hicks, K. Kemnitz, K. B. Eisenthal, and T. F. Heinz, J. Phys. Chem. 90:560 (1986); K. Bhattacharyya, E. V. Sitzman, and K. B. Eisenthal, J. Chem. Phys. 87:1442 (1987).

106. K. Bhattacharyya, E. V. Sitzman, and K. B. Eisenthal, J. Chem. Phys. 87:1442 (1987).

107. S. G. Grubb, M. W. Kim, Th. Rasing, and Y. R. Shen, Langmuir 4:452 (1988).

108. Th. Rasing, T. Stehlin, Y. R. Shen, M. W. Kim, and P. Valint, Jr., J. Chem. Phys. 89:3386 (1988).

109. M. W. Kim, S.-N. Liu, and T. C. Chung, Phys. Rev. Lett. 60:2745 (1988).

110. M. C. Goh, J. M. Hicks, K. Kemnitz, G. R. Pinto, K. Bhattacharyya, K. B. Eisenthal, and T. F. Heinz, J. Phys. Chem. 92:5074 (1988).

111. K. Kemnitz, K. Bhattacharyya, J. M. Hicks, G. R. Pinto, K. B. Eisenthal, and T. F. Heinz, Chem. Phys. Lett. 131:285 (1986).

112. Th. Rasing, Y. R. Shen, M. W. Kim, and S. Grubb, Phys. Rev. Lett. 55:2903 (1985).

113. Th. Rasing, Y. R. Shen, M. W. Kim, P. Valint, Jr., and J. Bock, Phys. Rev. A 31:537 (1985).

114. J. E. Sipe, J. Opt. Soc. Am. B 4:481 (1987).

115. E. Adler, Phys. Rev. 134:A728 (1964).

116. Hung Cheng and P. B. Miller, Phys. Rev. 134:A683 (1964).

117. S. S. Jha, Phys. Rev. 145:500 (1966).

118. P. Guyot-Sionnest, W. Chen, and Y. R. Shen, Phys. Rev. B. 33:8254 (1986).

119. P. Guyot-Sionnest and Y. R. Shen, Phys. Rev. B. 35: 4420 (1987).

120. T. F. Heinz, Ph.D. dissertation, University of California, Berkeley, 1982.

121. V. Mizrahi and J. E. Sipe, J. Opt. Soc. Am. B. 5:660 (1988).

122. J. E. Sipe, V. Mizrahi, and G. I. Stegeman, Phys. Rev. B. 35:9091 (1987).

123. P. Guyot-Sionnest and Y. R. Shen, Phys. Rev. B. 38: 7985 (1988).

124. M. Weber and A. Liebsch, Phys. Rev. B. 35:7411 (1987).

125. J. E. Sipe, V. C. Y. So, M. Fukui, and G. I. Stegeman, Solid State Comm. 34:523 (1980).

126. M. Corvi and W. L. Schaich, Phys. Rev. B. 33:3688 (1986).

127. O. Keller, J. Opt. Soc. Am. B. 2:367 (1985).

128. O. Keller, Phys. Rev. B. 31:5028 (1985).

129. O. Keller, Phys. Rev. B. 33:990 (1986).

130. O. Keller, Opt. Acta 33:673 (1986).

131. O. Keller and K. Pedersen, J. Phys. C: Solid State Phys. 19:3631 (1986).

132. O. Keller, Phys. Rev. B. 34:3883 (1986).

133. M. Weber and A. Liebsch, Phys. Rev. B. 37:1019 (1988).

134. A. Chizmeshya and E. Zaremba, Phys. Rev. B. 37:2805 (1988).

135. A. Leibsch, Phys. Rev. B. 36:7378 (1987).

136. G. C. Aers and J. E. Inglesfield, Electric fields and Ag(001) surface electronic structure, submitted.

137. D. M. Kolb, in Advances in Electrochemistry and Electro- chemical Engineering, Vol. II (H. Gerischer and C. W. Tobias, eds.), Wiley, New York, 1978, p. 125.

138. K.-M. Ho, C. L. Fu, S. H. Liu, D. M. Kolb, and G. Piazza, J. Electroanal. Chem. 150:235 (1983).

139. W. Boeck and D. M. Kolb, Surf. Sci. 118:613 (1982).

140. C. V. Shank, R. Yen, and C. Hirlimann, Phys. Rev. Lett. 51:900 (1983).

141. T. F. Heinz, G. Arjavalingam, M. T. Loy, and J. H. Glownia, in Proceedings of the XIV International Quantum Electronics Conference, June 9—13, 1986, San Francisco, CA. Paper THII1.

142. G. Arjavalingam, T. F. Heinz, and J. H. Glownia, in Ultrafast Phenomena V (G. R. Fleming and A. E. Siegman, eds.), Springer-Verlag, Berlin, 1986, p. 370.

143. H. W. K. Tom, G. D. Aumiller, and C. H. Brito-Cruz, Phys. Rev. Lett. 60:1438 (1988).

144. H. W. K. Tom, in Advances in Laser Science—II. AIP Conference Proceedings Series No. 180 (W. C. Stwalley, M. Lapp, and G. A. Kenney-Wallace, eds.), AIP, New York, 1987.

145. J. E. Sipe and G. I. Stegeman, in Surface Polaritons (V. M. Agranovich and D. L. Mills, eds.), North-Holland Publishing Company, New York, 1982, p. 661.

146. T. F. Heinz, M. M. T. Loy, and W. A. Thompson, Phys. Rev. Lett. 54:63 (1985).

147. T. F. Heinz, H. W. K. Tom, and Y. R. Shen, Laser Focus, May 1983.

148. G. T. Boyd, Y. R. Shen, and T. W. Hänsch, Opt. Lett. 11:97 (1986).

149. G. L. Richmond, Proceedings of 188th ACS National Meeting, Philadelphia, PA, 1984.

150. P. Guyot-Sionnest and A. Tadjeddine, J. Phys. Chem. 92:1 (1990).

151. M. G. Samant, M. F. Toney, G. L. Borges, L. Blum, and O. R. Melroy, J. Phys. Chem. 92:220 (1988).

152. M G. Samant, M. F. Toney, G. L. Borges, L. Blum, and O. R. Melroy, Surf. Sci. 193:129 (1988).

153. R. Sonnenfeld, J. Schneir, and P. K. Hansma, in Modern Aspects of Electrochemistry, Vol. 21, in press.

154. Dennis J. Trevor, C. E. D. Chidsey, and D. N. Loiacono, Phys. Rev. Lett. 62:929 (1989).

155. D. Guidotti, T. A. Driscoll, and H. J. Gerritsen, Solid State Comm. 46:337 (1983); T. A. Driscoll and D. Guidotti, Phys. Rev. B. 28:1171 (1983).

156. H. W. K. Tom, T. F. Heinz, and Y. R. Shen, Phys. Rev. Lett. 51:1983 (1983).

157. J. F. Nye, Physical Properties of Crystals: Their Representation by Tensors and Matrices, Claredon Press, Oxford, 1985.

158. H. W. K. Tom and G. D. Aumiller, Phys. Rev. B. 33:8818 (1986).

159. J. A. Litwin, J. E. Sipe, and H. M. van Driel, Phys. Rev. B. 31:5543 (1985).

160. H. W. K. Tom, Ph.D. dissertation, University of California, Berkeley, 1984.

161. J. E. Sipe, D. J. Moss, and H. M. van Driel, Phys. Rev. B. 35:1129 (1987).

162. R. Bradley, S. Arekat, S. D. Kevan, and G. L. Richmond, Chem. Phys. Lett., in press.

163. T. E. Furtak and D. W. Lynch, Phys. Rev. Lett. 35:960 (1975).

164. V. L. Shannon, D. A. Koos, and G. L. Richmond, unpublished results.

165. J. M. Robinson, G. A. Neff, and G. L. Richmond, in preparation.

166. D. M. Kolb, J. Phys. (Paris) 44:C10-137 (1983); D. M. Kolb and J. Schneider, Electrochim. Acta 31:929 (1986); J. Schneider and D. M. Kolb, Surf. Sci. 193:579 (1988).

167. B. E. Conway, Prog. in Surf. Sci. 16:1 (1984).

168. A. Bewick and B. Thomas, J. Electroanal. Chem. 65:911 (1975).

169. A. Bewick and B. Thomas, J. Electroanal. Chem. 85:329 (1977).

170. L. Laguren-Davidson, F. Lu, G. N. Salaita, and A. T. Hubbard, Langmuir 4:224 (1988).

171. K. Takayanagi, D. M. Kolb, K. Kamba, and G. Lehmphuhl, Surf. Sci. 100:407 (1980); A. Rolland, J. Bernadini, and M. B. Barthes-Labrousse, Surf. Sci. 143:579 (1984).

172. H. P. Myers, J. Phys. F: Metal Physics $\underline{3}$:1078 (1978);
     H. J. Liljenvall, A. G. Mathewson, and H. P. Meyers,
     Phil. Mag. $\underline{22}$:143 (1970).

173. J. M. Robinson and G. L. Richmond, Dynamics of thallium
     deposition on Cu(111), in preparation.

NEW DEVELOPMENTS IN
ELECTROCHEMICAL MASS SPECTROSCOPY

Barbara Bittins-Cattaneo,
Eduardo Cattaneo,
Peter Königshoven, and
Wolf Vielstich

Institute of Physical Chemistry
University of Bonn
Bonn, Federal Republic of Germany

I.   INTRODUCTION

Recently developed methods in electrochemical mass spectroscopy (MS) have become a powerful tool in electroanalysis and
electrochemical kinetics: Direct connection of the electrochemical cell with the MS system via a porous and hydrophobic electrode [1–12] offers a valuable extension of cyclic voltammetry.
In addition to current potential diagrams, one obtains multi-mass
intensity potential diagrams of volatile species consumed or produced at the electrode surface [8–14].   Processes at compact
electrodes or at single crystals can be followed in applying a
rotating electrode near the "membrane window" at the MS inlet
[10].   Examples are given from investigations in electrocatalysis,
batteries, sensors, and corrosion.

A.   Historical Overview and Basic Ideas

The possibility of measuring electrochemical reaction products
by connecting the electrochemical cell to a mass spectrometer
was first suggested by Bruckenstein [1].   The experimental
arrangement consisted in attaching the working electrode to a
porous membrane of small pore size and nonwetting properties,
which was used as the interface between the cell and the MS
inlet.   With the Bruckenstein device the delay of response for
the gaseous products entering the mass spectrometer was about
20 sec.

A similar setup as that of Bruckenstein was later constructed by Brockman and Anderson [15]. Working with a permeable 25-μm silicon rubber membrane, dimethylsulfoxide also could be used as a solvent. The formation of volatile products of the Kolbe oxidation of carboxylate ions in DMSO solvents and water was followed. The time resolution of the device was characterized by a half-life time between 10 and 100 sec.

For a quantitative analysis of gaseous electrolysis products, Kreysa et al. presented an experimental setup where the product gas is transferred from the electrochemical cell to the mass spectrometer via a carrier gas [7].

A considerable improvement of the method was achieved by Wolter and Heitbaum [8,9] through the use of a differential pumping system with turbomolecular pumps. This allows a fast transfer of products into the ionization chamber (the time response amounts to ca. 0.2 sec) and a rapid elimination of the gas collected. In this way potentiodynamic current and mass intensity profiles may be correlated without major distortions for scan rates up to 50 mV/sec. In order to emphasize the fact that the mass signal is instantaneously changing in the same way as the current does, the authors called this technique differential electrochemical mass spectroscopy (DEMS).

The original experimental procedure allows the analysis of electrochemical processes at porous noble metal electrodes prepared by applying a thin layer of a finely divided metal suspension on a nonwetting PTFE membrane. In optimizing the arrangement of the differential pumping device the signal response time was lowered to 50 msec in connection with a high sensitivity of the mass analyzer system [12,13].

A combination of the rotating disc technique with on-line mass spectroscopy on porous electrodes was suggested by Heitbaum [14] with the construction of a rotating inlet system. This approach can be used for the calculation of collection

efficiencies by correlating quantitatively the Faradaic current of
formation with the corresponding ion current at the detector
system.

As working electrodes the authors employed lacquer-type
electrodes as described above and a second type that is prepared by sputtering metal particles onto the PTFE membrane.
Whereas the roughness factor of the lacquer electrodes usually
ranges between 10 and 50, that of the sputtered electrodes lies
between 3 and 5. A disadvantage of the thin porous electrodes
is their mechanical instability against strong gas evolution reactions. This becomes a problem in the study of electrode reactions coupled with strong gas evolution.

B. Recent Developments

A principal extension of electrochemical on-line mass spectroscopy to the study of compact material as a working electrode
has been developed recently by Vielstich et al. [10]. Here the
solid catalyst material—for example, in form of a cylinder or a
disc—is positioned very close (ca. 0.3 mm) to the PTFE membrane of the mass spectrometer inlet. To achieve a radial component of mass flow, which transports the volatile products
toward the membrane, the cylindrical working electrode must be
in rotation. The radial component of mass flow increases with
rotation frequency. Depending on the actual system, at a certain frequency range the mass flow remains nearly constant and
a good signal-to-noise ratio of the mass signal is obtained.

Using this new device, investigations can be extended to
compact electrode material, single crystals, and to the case of
gas evolution reactions as well. An extension to the on-line
study of species in the bulk of the solution has been suggested
by Hambitzer and Heitbaum [16] connecting a setup for thermospray ionization with the working electrode of an electrochemical
cell (see Sec. VI.A).

In another method for studying the nature of strongly adsorbed particles on electrodes, the working electrode is taken out of the electrolyte and then immediately transferred to an UHV analysis chamber (see Sec. VI.B and Ref. 17).

II.   EXPERIMENTAL METHODS

The experimental setup used for the combination of cyclic voltammetry and on-line mass spectroscopy (DEMS) is shown in Fig. 1. It consists of an electrochemical cell attached to the inlet of a mass spectrometer system, which is evacuated differentially by two turbomolecular pumps (PA and PB). The electrodes of the cell are connected to a potentiostat driven by a function generator. The whole system is controlled through a computer.

A typical electrochemical cell used in many of our experiments is shown in Fig. 2. All cells used so far are based on the same working principle. They differ in volume and construction material, e.g., PTFE (Fig. 2), Kel-F [15], and glass [12].

A.   Flow Cell for On-Line Mass Spectroscopy

The cell of Fig. 2, constructed as a mini-flow cell, is made of PTFE or lucite and contains approximately 2 ml of solution. The solution can be added through an inlet tube made of glass ending just above the electrolyte level; the excess of electrolyte leaves the cell through an outlet. As the three-electrode system in the cell is always covered by electrolyte, this procedure keeps the working electrode under potential control during any change of electrolyte.

In order to avoid contamination with air, all solution containers are made of glass and are connected to the cell by glass tubes only. For the same reason, the inert gas used (99.996% Ar further purified through Oyxsorb, Messer-Griesheim) is led

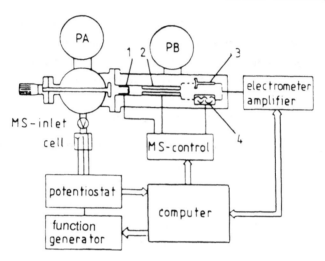

FIG. 1. Schematic diagram of the experimental setup used for
on-line mass spectroscopy: the electrochemical cell is connected
via an inlet system (see Figs. 2 and 3) to the mass spectrometer recipient pumped differentially by turbo pumps PA and PB;
volatile electrolysis products are ionized in the ionization chamber (1), analyzed with a quadrupole mass filter (2), and detected with a Faraday cup (3) or secondary electron multiplier
(4); the control of instruments and data acquisition are carried
out by a PC.

to the cell and solution containers through steel capillaries.
Gas outlets are sealed by water traps.

The bottom of the cell consists of a porous, nonwetting
PTFE membrane (Gore type S10570; thickness 75 μm, pore size
0.02 μm, porosity 50%) on top of a steel frit. The frit provides
the necessary mechanical stability against vacuum. A screw
coupling connects the cell to a stainless steel adapter supporting
the frit; a PTFE O-ring provides the hermetic seal.

The working electrode consists of a thin metal layer, e.g.,
a thin platinum layer prepared from a Pt suspension (DODUCO
PT5, Dürrwächter KG) of mean particle size 5 μm applied on

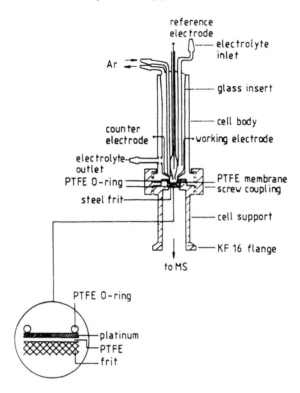

FIG. 2.   Flow cell with low electrolyte volume used for on-line mass spectroscopic experiments with porous metal electrodes; the cell body is made out of PTFE or lucite.

the PTFE membrane.   A Pt wire (ϕ = 0.5 mm, the tip molten to a sphere) pressed gently against the Pt layer provides the connection to the potentiostat.

Prior to experimental use the electrodes are activated by cycling (v = 300 mV/sec) during several hours between the onset potentials for H_2 and O_2 evolution in 0.5 M H_2SO_4, while changing the electrolyte several times.   This procedure is repeated until the surface layer diagram shows the usual shape, indicating a clean Pt surface without traces of lacquer solvent.

Depending on the preparation, the real area obtained varies
between 10 and 50 cm^2, the geometric area amounting to ca.
0.5 cm^2.

The potentials are usually measured versus a reversible
hydrogen electrode in the same solution. A palladium wire or
net, platinized and charged with hydrogen, is especially suitable. A platinum wire or platinum sheet usually serves as
counterelectrode.

B.   Rotating Porous Electrode as Inlet System

An experimental setup combining a rotating disc device with on-
line mass spectroscopy has recently been described [14]. In
this case the PTFE membrane with its porous catalyst layer
(conducting Pt lacquer or sputtered Pt particles) is attached to
a rotating feed-through connected to the mass spectrometer.
This variation of the original technique is suitable for studying
diffusion controlled processes under stationary conditions. It
permits the separation of the diffusion contribution from the
total current. The method was tested by studying the hydrogen
evolution reaction in perchloric acid for different isotopes. Due
to the relatively poor adherance of the catalytic layer to the
membrane, the current for gas evolution has to be limited to a
few milliamperes.

C.   MS Inlet for Compact Electrode Material

Porous metal layer electrodes do not allow the investigation of
electrode reactions connected with strong gas evolution. This
is the case, for example, when studying CO_2 reduction to methane and ethene on Cu electrodes in $KHCO_3$ solution. The production of hydrocarbons occurs in this case at negative potentials (-1.7 V vs. SCE) where large amounts of hydrogen in
form of gas bubbles are produced. The strong bubbling causes
destruction of any thin porous electrode in a few minutes.

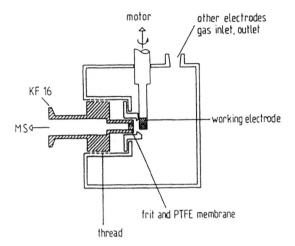

FIG. 3. Electrochemical cell used for on-line mass spectroscopic experiments with compact electrode material in the form of a rotating cylinder. (From Ref. 10, with permission of Pergamon Press.)

To overcome this difficulty, a new type of cell using compact electrode material was constructed. In Fig. 3 the working electrode consists of a cylindrical rod of copper or glassy carbon positioned close to the PTFE membrane (ca. 0.3 mm). A rotation of the cylinder is necessary in order to achieve a radial component of mass flow, which transports volatile products towards the membrane (i.e., towards the inlet of the mass spectrometer). The radial component of mass flow increases with rotation frequency. With our setup a good signal-to-noise ratio is obtained between 40 and 60 Hz (see below and Figs. 15 and 16).

Often other geometries of rotating electrodes are suitable, e.g., in the case of single crystals, a disc-shaped electrode is preferred.

D. Instrumentation

The electrochemical cell is attached to the mass spectrometer system via the KF16 flange of the inlet valve (Fig. 1). All connecting parts of the recipient are made of (electropolished) stainless steel V2A, thus for evaporation of impurities, the device can be heated up to a temperature of 120°C.

The extended view of Fig. 4 shows two vacuum chambers: (A), evacuated through the turbo pump PA (Balzers TPU 170) with a working pressure of typical 10^{-1} mbar, and (B), differentially pumped through a second turbo pump PB (Balzers TPU 050), placed between the ionization chamber and the quadrupole

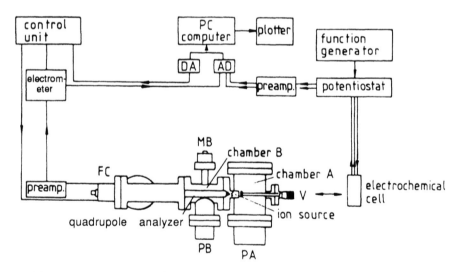

FIG. 4. Periphery of the on-line mass spectroscopy setup; valve V adjusts the flow of product gas in chamber B (pressure measured with cold cathode gauge MB) coming from inlet chamber A; the recipient is differentially pumped by turbo pumps PA and PB; a PC controls the electrometer and stores the electrometer and potentiostat output.

rods, where a much smaller working pressure of less than 10^{-4} mbar is reached. An all-metal stainless steel valve ending in a plate which is driven through a 150 mm long axis by a micrometer screw provides a very good metal-to-metal seal to isolate chamber B from chamber A. This manually operated valve also allows a fine setting of the appropriate working pressure in chamber B, i.e., the vacuum required for the ion source, the secondary electron multiplier etc. When the valve is closed, the pressure in chamber B amounts to 10^{-8} mbar, making possible a manipulation of the cell without switching off the emission of the spectrometer. This speeds up the experimental procedure and prolongs the life of the ion source filament.

The volatile products formed at the working electrode and the solvent vapor (e.g., water, propylene carbonate, or dimethylsulfoxide) which reach the ionization chamber in volume (B) are ionized by electron bombardment. The ions produced are accelerated towards the mass filter (quadrupole rods), but only the ions with the desired m/e ratio reach the detector. The m/e ratio is selected manually or with the help of a computer and the control unit Balzers QMG 311 (Fig. 4). Two different detector systems may be used: a Faraday-cup and a SEM (secondary electron multiplier) placed perpendicular to the main axis. The mass signal from the Faraday cup is preamplified and measured with an electrometer.

The periphery of the quadrupole mass spectrometer with connections is shown schematically in Fig. 4. The automatic sampling of data, which allows multiple ion detection, is performed using a personal computer (Commodore PC20-III) with a built-in AD/DA interface (AD-Elektronik). The AD/DA interface consists of two 12-bit DA converters (conversion time 15 μsec) configured to supply an output voltage between 0 and 10 V in two channels, and one 12-bit AD converter multiplexed to 16 differential input channels operating between 0 and 10 V (conversion time 40 μsec including multiplexing). The voltage resolution for AD and DA conversions amounts to 2.44 mV.

One of the DA channels is used for selecting the m/e value to be measured. For this purpose, the QMG 311 control unit provides an analog input (0 to 10 V) permitting one to set, from the computer, m/e values ranging between 0 and 100 or 0 and 300 AMU with a resolution of 0.024 and 0.073 AMU (AMU = atomic mass unit), respectively. The corresponding mass signal is read by one of the 16 AD channels.

The electrochemical measurements are performed with a potentiostat driven by a function generator. Current and potential output of the potentiostat are first amplified (Fig. 4) and then read by two of the remaining AD channels.

III.   FEATURES AND LIMITATIONS OF
ON-LINE MS TECHNIQUES
A.   Mass Intensity Potential Voltammograms

An electrochemical cell mounted to a mass spectrometer allows the simultaneous registration of cyclic current potential and mass intensity potential voltammograms. When comparing the intensities of different masses quantitatively, one has to keep in mind that ionization probabilities and fragmentation of different species can be quite different.

As an example, Fig. 5 shows results obtained for the anodic oxidation of ethanol on platinum in acid solution [18]. Figure 5a gives the usual current potential diagram covering the potential region of interest between beginning hydrogen and oxygen evolution. In Fig. 5b mass signals for volatile species formed during the oxidation process are presented. Besides a heavy CO_2 production (the respective signals m/e = 44 and 45 ($^{12}CO_2$ and $^{13}CO_2$) are not shown in Fig. 5b), potential dependent mass signals for m/e = 15 and 30 can be detected. It has to be

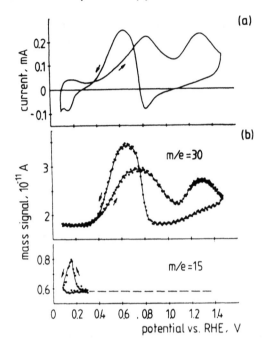

FIG. 5. Oxidation of ethanol on a porous thin layer platinum electrode (roughness factor ca. 50) in a 0.01 M $^{12}CH_3^{13}CH_2OH$ + 0.05 M H_2SO_4 solution, v = 10 mV/sec (a) Current potential and (b) mass intensity potential voltammograms (m/e = 30 acetaldehyde, m/e = 15 methane). (From Ref. 18, with permission of Bunsengesellschaft.)

noted that the ethanol used was isotope labeled, 0.01 M $^{12}CH_3^{13}CH_2OH$ in 0.05 M H_2SO_4 solution. We find a potential dependent mass signal for m/e = 15 which represents the M1 signal of $^{12}CH_4$ methane formed by the cleavage of one H atom in the ionization source. A high intensity mass signal is observed for mass m/e = 30 (main fragment of $^{12}CH_3^{13}CHO$ acetaldehyde). Simultaneously masses m/e = 29 (main fragment of $^{12}C^{13}CH_6$ ethane), m/e = 32 (main fragment of $^{13}CH_3OH$ and $^{12}CH_3^{13}CH_2OH$ ethanol) were recorded. For masses m/e = 29 and 31, no potential dependence above the respective base intensity could be detected. The same holds for $m/e = 32$ except for the clearly visible effect of ethanol consumption at the electrode surface (see Ref. 19).

Without discussing details, Fig. 5 clearly demonstrates the essential increase in information by recording simultaneously current and mass signals in the potentiodynamic experiment.

Another example is given in Fig. 6. The upper curve (a) shows the current potential scan of 1.3 ^{15}N-urea Krebs-Ringer solution at porous platinum [20]. The oxidation of urea is a matter meriting special attention.*

The current potential diagram of curve a in Fig. 6 (the dashed curve was obtained in supporting electrolyte) offers only little information about the electrooxidation of urea; e.g., the voltammogram does not show a positive current during the negative scan. But it is obvious from the MS results that an oxidation process occurs during the cathodic scan producing N_2 (curve c) and CO_2 (not shown in Fig. 6). In addition, three nitrogen oxides have been discovered as volatile oxidation products. Nitric oxide, NO, can in principle be investigated by following the signal $m/e = 30$ for unlabelled NO or 31 for ^{15}N-labeled NO. These signals, however, can also be produced by NO_2 and N_2O. We have monitored the signal $m/e = 31$ in ^{15}N-urea solutions, during the potential scan (curve b).

Obviously some nitrogen oxides are formed positive to 1 V vs. RHE during the positive scan and near 0.9 V during the negative scan.

---

*The electrochemical behavior of urea must be considered in the development of glucose-based devices, such as sensors or bio-fuel cells for pacemakers and artificial hearts. In this case urea acts as a co-reactant and can affect the performance of such devices.

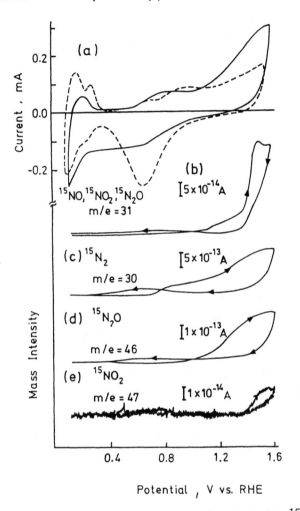

FIG. 6. Potentiodynamic investigation of 0.03 M 1.3 ^{15}N-urea in Krebs-Ringer solution on a porous thin layer platinum electrode, real Pt area 35 cm^2, v = 10 mV/sec. (a) Current-potential voltammogram, dashed line obtained from the scan in supporting electrolyte; (b) mass intensity potential scans associated with NO, NO_2, and N_2O, (c) N_2, (d) N_2O, and (e) NO_2, respectively. (From Ref. 20, with permission of Pergamon Press.)

Curve (d) shows the contribution of $^{15}N_2O$ via the voltammogram of m/e = 46. The contribution from CO_2 to the mass
signal m/e = 46 was found to be very small [20]. N_2O appears
at potentials above 0.8 V during the anodic scan; in the cathodic
scan N_2O production can be observed at very low potentials.

NO$_2$ can be followed without interferences using the m/e-
signal 47 in labeled ^{15}N-urea. This species is shown to occur
above 1.2 V (curve e).

B.   Time Resolution of On-Line Mass Spectroscopy

In studying the voltammograms of Figs. 5 and 6, which are obtained using a scan rate of 10 mV/sec, it is a question up to
which scan rate the mass signal will be resolved sufficiently.
Figure 7 demonstrates that the mass signal resolution easily
follows even the structured current time (potential) relationship
during the oxidation of formic acid at platinum in a heavily
purged solution. The m/e = 44 signal for CO_2 (product of
HCOOH oxidation) follows precisely the current signals as function of time or potential, in distinctive potential regions.

Under favorable conditions the time shift between the production of a volatile species at the electrode surface and the
respective mass signal is only 50 msec. In such cases a successful recording of voltammograms up to scan rates above 100
mV/sec is possible.

C.   Sensitivity of the Mass Signal Voltammogram

In many investigations the mass intensity signal offers a more
sensitive insight to surface conditions of electrodes than the
current potential voltammogram does. In the example shown in
Fig. 8 the cleanness of the interface electrode/electrolyte is
studied after experiments with $^{13}CH_3OH$ containing electrolyte
at a porous platinum electrode in H_2SO_4 solution. The cyclic

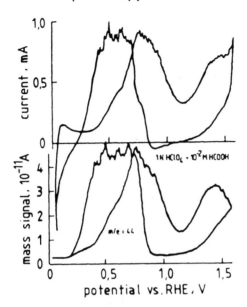

FIG. 7. Time resolution of a cyclic on-line mass intensity potential voltammogram; simultaneously taken current potential and mass intensity potential scans for the oxidation of 10^{-2} M HCOOH in 1 M $HClO_4$ on a porous Pt layer as in Figs. 5 and 6, but using a sweep rate of 12.5 mV/sec and nitrogen purging, m/e = 44. Signal shows evolution of CO_2.

current potential voltammogram taken after exchange of electrolyte by a fresh 1 M H_2SO_4 solution does not show any defect and no hint of any impurity at the electrode surface. However, the simultaneous registration of $^{13}CO_2$ via m/e = 45 clearly proves that residual species from the ^{13}C-methanol experiment are still adsorbed at the porous platinum surface. Experience shows that coverages lower than one percent of a monolayer can be detected.

In the case of compact material as working electrode, e.g., rotating disc or cylinder, only a minor part of the species formed is transferred in the MS via the inlet membrane. But

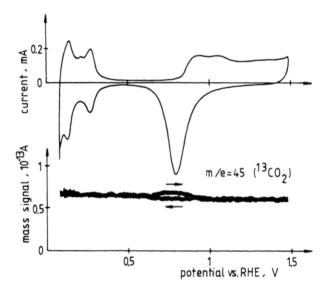

FIG. 8. Impurity check of a porous platinum electrode previously used for experiments with $^{13}CH_3OH$ by on-line mass
spectroscopy; simultaneously taken current potential scan using
1 M H_2SO_4 as electrolyte and mass intensity potential scan of
$^{13}CO_2$ (m/e = 45); real Pt surface, 35 cm^2; sweep rate, 10 mV/
sec.

the signals obtained are still sufficient to follow the oxidation of
adsorbates in the order of a monolayer.

IV.   STUDY OF ADSORBATES AND
CATALYTIC PROCESSES

In this section three examples are given for the application of
on-line MS to the study of poisoning adsorbates like carbon
monoxide.   Special attention is given to the electrocatalysis of
ethanol and methanol.

A.   Carbon Monoxide as Adsorbate, a
Calibration Experiment

A quantitative relation between the charges under the current and mass intensity signals can be established, as suggested by Wolter and Heitbaum [21].   The magnitude of the mass intensity response does not only depend on the electrochemical properties of the system under study, but also on the permeability of the electrode to the volatile products in addition to mass spectrometer parameters.   The experimental procedure is as follows:

1.   After activation of the porous platinum electrode in 0.5 M H_2SO_4, the potential is stopped in the cathodic scan at E_{ad} = +0.2 V vs. RHE.
2.   CO is supplied during 2 min and afterwards removed by 12 min Ar bubbling; during the adsorption the current transient is recorded.
3.   Starting at E_{ad} = +0.2 V vs. RHE, the potential scan is continued in cathodic direction to 0.05 V and then anodic up to 1.5 V; during this procedure current potential and mass intensity potential curves for m/e = 44 (CO_2) are recorded simultaneously.

The respective diagrams are given in Fig. 9.   The mass intensity MI of the electrochemically generated product is directly proportional to the Faradaic current I of its formation:

$$MI = K \frac{1}{nF} AI \tag{1}$$

where K is a constant including all possible parameters of the MS device which are not specific for a given electrochemical process, n is the number of electrons necessary to produce one molecule of product, F is the Faraday constant, and A is the current efficiency of the electrode process studied.

Bittins-Cattaneo, et al.

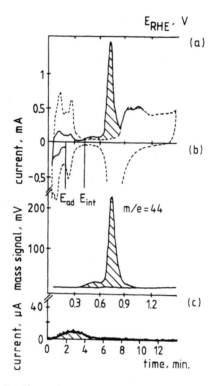

FIG. 9. (a) Cyclic voltammogram and (b) mass signal m/e = 44 (CO_2) for the electrodesorption of preadsorbed CO on a porous platinum electrode in 0.5 M H_2SO_4 at E_{ad} = +0.2 V vs. RHE, according to Wolter and Heitbaum [21]; v = 12.5 mV/sec (E_{int} = 0.375 V). Dashed line: surface layer diagram in 0.5 M H_2SO_4; for meaning of the hatched areas see text; (c) Current transient during adsorption of CO.

Equation (1) can be written in terms of the charges Q_I and Q_{MI} obtained by integration of current and mass signals as function of time, respectively:

$$Q_{MI} = K^0 \frac{1}{n} A Q_I \tag{2}$$

where K^0 also includes the Faraday constant.

Charge Q_{MI} represents the integral under the mass signal potential curve (see Fig. 9b), while Q_I is the total charge that flows during the adsorption of CO (see Fig. 9c) and then during the oxidation of CO to the product CO_2 (see Fig. 9a, hatched area).

The above calculation is correct, if CO_2 is the only oxidation product, i.e., no desorption intermediates occur and the adsorbate is completely oxidized and removed after the complete potential cycle. These conditions are fulfilled with an adsorption potential of $E_{ad} = 0.2$ V. Consequently, the current efficiency A equals 1 and the number of electrons for the overall reaction amounts to n = 2. Thus one obtains:

$$K^0 = \frac{2 Q_{MI}}{Q_I} \tag{3}$$

As the integration procedure starts and ends at the same potential E_{int} with an uncovered electrode surface (see Fig. 9a), all capacitive and pseudocapacitive effects cancel.

Using the procedure described above one determines constant K^0 for the actual electrode with a well-known electrochemical reaction. In principle it is possible now to investigate the stoichiometry of other adsorbates that can be oxidized to carbon dioxide.

B.   Ethanol Adsorbates on Platinum

As an example for the elucidation of an adsorbate composition,
the results on the adsorbate originating from ethanol oxidation
on platinum in acid solution are given in Fig. 10.

The adsorbate is formed on a porous platinum electrode in

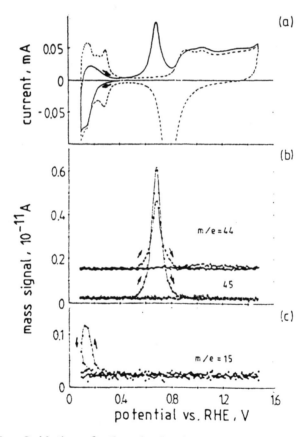

FIG. 10.   Oxidation of ethanol adsorbate on a porous platinum
electrode; ethanol adsorbate formed in flow cell technique from
a 0.01 M $^{12}CH_3{}^{13}CH_2OH$ + 0.05 M H_2SO_4 solution, E_{ad} = 0.4 V
vs. RHE; t_{ad} = 200 sec.   (a) Current (dashed line: base electrolyte) and (b) mass signals for m/e = 44 ($^{12}CO_2$) and 45
($^{13}CO_2$), (c) mass signal for m/e = 15 (methane).   (From Ref.
18, with permission of Bunsengesellschaft.)

a flow cell procedure from a 0.01 M $^{12}CH_3^{13}CH_2OH$ + 0.05 M H_2SO_4 solution at 0.4 V vs. RHE during 200 sec (change of solution after alcohol adsorption to base electrolyte under potential control [12]). Isotope-labeled ethanol was used in order to distinguish between the oxidation products of the alcoholic and the methyl group of the ethanol molecule.

In Fig. 10b of the figure, two different kinds of CO_2, namely $^{12}CO_2$ (m/e = 44) and $^{13}CO_2$ (m/e = 45), are obtained showing the same peak potential. Mass m/e = 15 (curve c) represents the M1 signal of $^{12}CH_4$ methane formed by the cleavage of one H atom in the ionization source of the mass spectrometer.

From these experimental results, one can conclude that the ethanol molecule adsorbs on platinum in two different ways. One adsorption path leads to the dissociation of the ethanol molecule into COH or CO species, which are finally oxidized to CO_2 at the same potential. On the other hand, the production of methane in the reductive scan supports the existence of another kind of adsorbate. As methane gas itself does not adsorb on platinum, the methane produced out of the adsorbate can only originate from the decomposition of a 2C-surface species. Comparing the areas under the respective mass signals shows that only a minor part of the adsorbed ethanol species keep their C–C bond intact.

The next example demonstrates the effect of tin ions acting as co-catalyst in the oxidation process of methanol adsorbate on platinum. Figure 11 shows the strong catalytic enhancement in current and CO_2 signal produced by the addition of Sn(II) ions during the oxidation process of methanol adsorbate on platinum (CO_2 is known to be the only product of methanol adsorbate oxidation). This effect can be observed at oxidation potentials as low as 0.25 V, whereas the oxidation of the methanol adsorbate on pure platinum only starts at potentials above 0.45 V.

After 15 min of interaction with tin (addition at E = 0.475 V), the remaining methanol adsorbate was oxidized within a potential

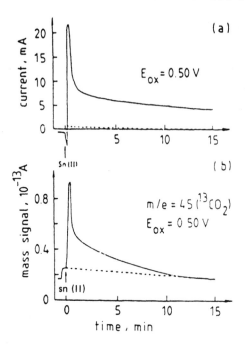

FIG. 11. Mass spectroscopic detection of CO_2 during methanol adsorbate oxidation on a porous platinum electrode; methanol adsorbate formed in flow cell technique from a 0.01 M $^{13}CH_3OH$ + 0.5 M H_2SO_4 solution at 0.4 V vs. RHE during 11 min; after electrolyte exchange the potential was stepped to E_{ox} = 0.5 V. Dashed line: without tin; solid line: after injection of Sn(II) (4 × 10^{-4} M $SnSO_4$ + 0.5 M H_2SO_4.) (a) Current as function of time, (b) mass signal as function of time. (From Ref. 22, with permission of Elsevier Scientific Publishing.)

scan, as shown in Fig. 12. Because of the overlapping oxidation currents originating from the methanol adsorbate and the adsorbed tin ions, the processes cannot be separated out of the electrochemical data only. However, on-line MS allows the observation of the organic residue alone by following the CO_2 mass signal. The large shift of the onset of methanol adsorbate oxidation to more cathodic potentials due to adsorbed Sn(II) species

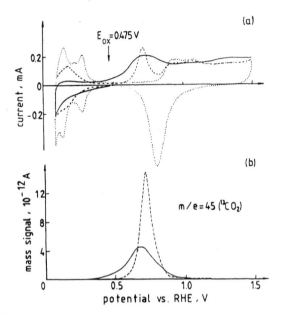

FIG. 12.   (a) Current and (b) mass intensity voltammograms
for methanol adsorbate oxidation on a porous platinum electrode
(dashed line: without tin; solid line: with 4×10^{-4} M SnSO$_4$
in solution; dotted line: base electrolyte 0.5 M H$_2$SO$_4$).   Methanol adsorbate formed in flow cell technique from 0.01 M
^{13}CH$_3$OH + 0.5 M H$_2$SO$_4$ solution at 0.4 V vs. RHE during 11
min followed by 15 min interaction with SnSO$_4$ addition at 0.475
V (as in Fig. 11).   (From Ref. 22, with permission of Elsevier
Scientific Publishing.)

is evident from Fig. 12 (presence of adsorbed tin, solid line;
pure platinum, dashed line).

V.   APPLICATIONS IN BATTERIES, CORROSION, ELECTRO-
SYNTHESIS, AND WATER ELECTROLYSIS

A.   Gas Evolution in the Lead Acid Cell

The study of hydrogen and oxygen evolution on lead using on-
line mass spectroscopy is particularly interesting for the investigation of processes occurring in lead batteries [23].   It is

well known that by charging a lead accumulator above an operating voltage of about 2.4 V the electrolytic decomposition of water
begins. H_2 and O_2 evolve on the negative and positive plates,
respectively. In order to reduce the hydrogen evolution on the
negative plates (Pb side), lead alloys like PbCa and PbCaSn are
used. These materials show a higher H_2 overpotential than the
traditional lead antimony alloy.

In conventional potentiodynamical experiments on electrode
materials the current is due to different Faradaic processes occurring simultaneously on the electrode surface. On-line mass
spectroscopy provides a direct method for monitoring the gas
evolution separately as function of potential. Indeed, the sensitivity of the mass spectroscopic system permits the detection
of very small quantities of gas, which leads to an accurate determination of onset potentials of gas evolution.

Figure 13 shows a cyclic voltammogram for a porous Pb
electrode in 0.5 M H_2SO_4 together with mass signal m/e = 2 for
hydrogen (H_2). According to the increase in cathodic current
H_2 evolution starts at −0.9 V vs. RHE. The mass signal clearly
shows hydrogen formation near −0.7 V.

B.   Oxidation of Propylene Carbonate

The on-line MS technique using a porous PTFE membrane as
support for different metal layers (e.g., Pt, Pb, Pd, etc.) is
possible due to the nonwetting properties of PTFE when working
with aqueous solutions. This method can be applied in the case
of widely used organic solvents like propylene carbonate (PC)
or dimethylsulfoxide (DMSO) [24]. PC is a well-known solvent
component for many types of lithium batteries and therefore the
study of its chemical stability as function of electrode material,
potential, and electrolyte composition is of interest.

Figure 14 shows the oxidative formation of CO_2 out of a

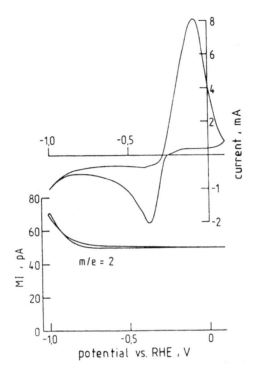

FIG. 13.   On-line mass spectroscopy during cyclic voltammetry
(potential scan 50 mV/sec) of a porous Pb electrode (Barton
lead) in 1 N H_2SO_4, Ar atmosphere and room temperature.   The
rise of the m/e = 2 mass signal above its ground level at potentials below −0.75 V vs. RHE indicates the onset of hydrogen
evolution.   (From Ref. 23, with permission.)

0.5 M $LiClO_4$ solution in PC on a porous platinum electrode by
stepping the potential between 2 and 4 V vs. Li [25].   The
rapid decrease of the CO_2 mass signal with time is probably due
to a "passivation-like" layer formed by PC oxidation products.
This layer blocks the further oxidation of PC and the CO_2 production diminishes.   These findings were complemented using an
in situ infrared technique (FTIR) [26] which also permits one

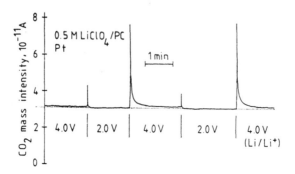

FIG. 14. Time dependence of the CO_2 (m/e = 44) mass signal as response to regular potential steps between 2 and 4 V vs. Li/Li$^+$ on a porous platinum electrode in 0.5 M LiClO$_4$ in propylene carbonate (PC) according to Ref. 25. The CO_2 production, shown by the steep increase of the m/e = 44 mass signal, is due to the oxidation of propylene carbonate at positive potentials. Further oxidation of PC is blocked by a layer formed out from its oxidation products. This leads to a strong decrease of the CO_2 signal after reaching its maximum.

to monitor CO_2 evolution in response to potential steps. A small amount of CO_2 is evolved also (see Fig. 14) when stepping the potential from 4 back to 2 V. This may be caused by CO_2 desorbed from the surface layer due to some potential dependent structural change [25]. In any case the on-line technique allows the immediate control of corrosion processes as soon as volatile products are formed.

C. Electroreduction of CO_2

The examples given so far are studies on porous electrodes. As mentioned above, thin porous metal layers cannot be used for studying processes with strong gas evolution. Such an electrochemical process must be studied using compact electrode material like electrodes for technical applications (batteries, electrolytic cells). The cell using a rotating cylinder electrode overcomes these limitations (Fig. 3). As an example, the electroreduction of CO_2 on copper in CO_2-saturated $KHCO_3$ solution is discussed. This reaction is accompanied with a strong hydrogen evolution [10].

Figure 15 shows the mass signals m/e = 15 (methane) and m/e = 26 (62% fragment of ethene) as a function of potential for a copper cylinder electrode (Ventron 32 mm length; 8 mm diameter) in CO_2 saturated 0.5 M $KHCO_3$ solution during a potential scan of 5 mV/sec. Both masses are measured successively within time intervals of 200 msec. The increase of both mass signals above the respective base lines below −1.7 V vs. SCE

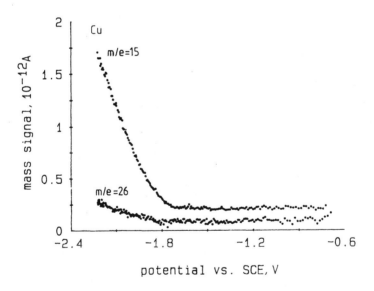

FIG. 15. Formation of methane (m/e = 15) and ethene (m/e = 26) on a rotating copper electrode (cell as in Fig. 3) as function of potential in CO_2-saturated 0.5 M $KHCO_3$ solution at room temperature (according to Ref. 10). Potential scan, 5 mV/sec; rotation frequency, 50 Hz; electrode surface, 7.4 cm^2.

demonstrates clearly the potential dependence of the methane
and ethene production.

The time dependence for the methane and ethene mass signals together with the corresponding current at a potential of
−2.2 V vs. SCE is shown in Fig. 16 for copper electrodeposited
on a glassy carbon cylinder electrode (Si gradur K 40 mm
length, 7 mm diameter). The decrease of intensity for both
mass signals after some minutes is obviously due to graphite
deposition on the electrode surface [27].

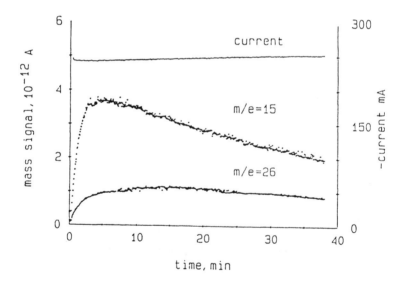

FIG. 16. Formation of methane (m/e = 15) and ethene (m/e =
26) at copper, electrodeposited on a rotating glassy carbon
electrode, during a transient experiment at −2.2 V vs. SCE in
CO_2-saturated 0.5 M $KHCO_3$ solution at room temperature (according to Ref. 10). Rotation frequency, 50 Hz; electrode surface, 7.4 cm^2. The decrease of both mass signals with time
after reaching a maximum is due to the slow poisoning of the
electrode resulting from the formation of a graphite layer on
its surface [27].

The advantage of on-line mass spectroscopy when compared with gaschromatographic methods for studying CO_2 electroreduction [27,28] is given by simultaneous monitoring of several masses; e.g., H_2 (m/e = 2), CO_2 (low intensity fragment m/e = 29), methane (fragment m/e = 15), and ethene (fragments m/e = 26 and 27) mass signals can be recorded.

The response time for detecting a mass signal after a change in potential depends critically on the distance between cylinder and PTFE membrane (about 0.3 mm, see Fig. 3). For example, in 0.5 M $KHCO_3$ solution, the mass signal for hydrogen is observed ca. 400/msec after a potential step from −0.7 to −1.0 V vs. SCE.

D. Oxygen Evolution on Metal Electrodes

Until now little work has been done on the study of the oxygen evolution reaction using on-line mass spectroscopy. Interest has focused on the elucidation of the role the oxide layer plays in the O_2 evolution mechanism on Pt, Ru, and RuO_2 in acid solution.

Using isotope labeling it was found that on platinum the oxide layer does not take part in O_2 evolution [29]. This result was obtained by forming an ^{18}O-oxide layer on a porous Pt electrode followed by the evolution of O_2 out of a $H_2^{16}O$-containing sulfuric acid solution. Only $^{16}O^{16}O$ could be observed mass spectroscopically, although the ^{18}O-oxide was still present on the electrode surface.

Similar experiments were repeated with rf-sputtered Ru and RuO_2 film electrodes [30]. With on-line mass spectroscopy it could be shown that $^{16}O^{18}O$ evolves out of a sulfuric acid solution containing $H_2^{16}O$ on a Ru oxide layer previously formed in $H_2^{18}O$. Thus, it was concluded that—in contrast to the case of Pt electrodes—for Ru and RuO_2 electrodes oxygen evolution occurs via the formation and decomposition of Ru oxides.

VI.   RECENT DEVELOPMENTS

A.   Electrochemical Thermospray Mass
Spectroscopy (ETMS)

In order to achieve a coupling between liquid chromatrography
and mass spectroscopy for the investigation of species in the
bulk of solutions, Blackley and Vestal [31] invented a thermospray ionization technique.   The total LC effluent is vaporized
into the ion source of a mass spectrometer with flow rates of
some cm^3/min.

Hambitzer and Heitbaum [16] successfully extended the procedure for the detection of electrogenerated products in solution.   The experimental setup is illustrated in Fig. 17.   An appropriate electrochemical cell is directly connected to the thermospray device.   The electrolyte is forced by pressure from
the working electrode into the heated capillary tube of a thermospray ion source.   A liquid N_2 buffer together with a rotary
vane pump condenses the vaporized electrolyte.   A Pt wire of
0.05 mm diameter and 600 mm length in the form of a helix (0.9
mm diameter and 4 mm length) is used as working electrode.
The helix is inserted into a PTFE tube in such a way that the
electrolyte passes the working electrode immediately before entering the mass spectrometer unit.   Studying the electrooxidation of dimethylaniline, a dead time of 9 sec was observed between the formation of species and the mass signal response
[16].

By improving the hydrodynamics of the central part of the
Thermospray unit essentially, Volk, Yost, and Brajter-Toth
succeeded in lowering the delay time down to only 500 msec in
their on line analysis [32].   Figure 18 shows the respective
voltammograms (200 mV/sec) for the electrochemical reaction of
uric acid on graphite.   Mass intensities for $m/e = 167$ $((M - H)^-$
of uric acid) as well as for products $m/e = 183$ (imine alcohol)
and $m/e = 157$ (allantoine) are given.

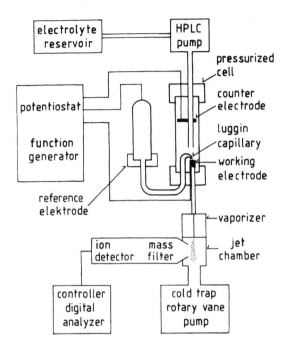

FIG. 17.   Schematic diagram of the experimental setup for electrochemical thermospray mass spectrometry (ETMS) according to Hambitzer and Heitbaum [16].

B.   Electrochemical Thermal Desorption
Mass Spectroscopy (ECTDMS)

The recently developed ECTDMS technique is especially suitable for the study of adsorbates on electrodes [17,33].   It is, of course, not an in situ procedure.   After completion of the adsorption in the electrochemical cell the working electrode is transferred to an UHV analysis chamber, connected to a quadrupole mass spectrometer.

A scheme of the experimental setup is shown in Fig. 19. The electrode is first transferred to a lock chamber where the liquid film covering its surface is eliminated by pumping to 10^{-6}

Bittins-Cattaneo, et al.

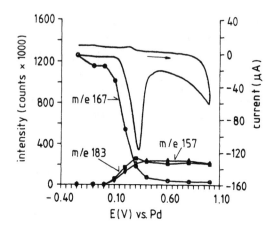

FIG. 18.   Cyclic current/potential and mass intensity/potential
voltammograms obtained via ETMS technique for uric acid on
rough pyrolytic graphite, according to Volk et al. [32], flow
rate of 0.1 M ammonium acetate mobile phase 2.0 cm^3/min; tip
temperature 240°C; source temperature 290°C; 300 µM uric acid;
palladium reference electrode; 200 mV/sec.

mbar with a turbo molecular pumping system, and then to the
UHV (p = 10^{-8}–10^{-9} m bar).   Here the desorption is achieved
by heating the electrode with the radiation of white light.   The
desorbing particles are focused onto a grid ion source of a
quadrupole mass spectrometer (Balzers QMG 112), supplied with
a multiple ion detection system (Balzers QMG 420), so that up
to 16 masses m/e can be recorded simultaneously.

The pumping system consists of turbo-molecular pumps (Balzers TPU 110 and 170).   In order to reduce the H$_2$ partial pressure, oil diffusion pumps are used as additional booster pumps.

In principle, the analysis of molecules, ions, and adsorbed
intermediates is possible if they survive the emersion (no potential control) and UHV conditions (elimination of most of solvent).
The use of ex situ methods for the analysis of submonolayer
quantities of oxygen-sensitive substances requires an extremely

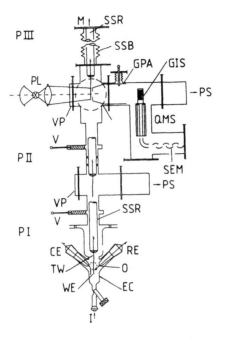

FIG. 19.   Experimental setup for ECTDMS. PI: electrode in the electrochemical cell; PII: electrode in the lock chamber; PIII: electrode in the UHV analysis position. CE: counterelectrode; EC: electrochemical cell; GES: grid ion source; GMS: glass to metal seal; GPA: gold-plated aperture; I: inlet for electrolyte; M: motor; O: outlet for electrolyte; PL: projector lamp with optical system; PS: pumping system; QMS: quadrupole mass spectrometer with secondary electron multiplier (SEM); RE: reference electrode; SSB: stainless steel bellows; SSR: stainless steel rod; TW: thermocouple wires; V: valve; VP: viewport; WE: working electrode; WS: wobble stick.

inert atmosphere when the electrode is emersed. In order to check whether a given adsorbate survives the experimental conditions, a control experiment must be done.

After adsorption of CO and solution exchange with pure base electrolyte, the oxidation of adsorbed CO during a potential scan is followed (see Fig. 20). In a second run after adsorption

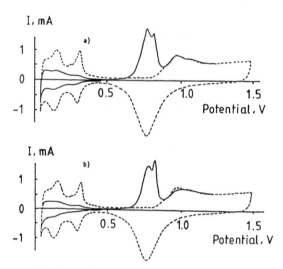

FIG. 20. Oxidation of CO adsorbate on platinum in 0.05 M
H_2SO_4 solution (a) in a flow cell experiment and (b) after
transfer to UHV analysis chamber and reimmersion into base
electrolyte, scan rate 62.5 mV/sec. Dotted line: voltammogram
without CO.

of CO, the electrode is immersed and transferred to the UHV
chamber in the same way as in the normal ECTDMS procedure.
After 10–15 min under UHV conditions, the electrode is transferred back to the cell and reimmersed. Again a potential scan
is applied to oxidize the adsorbate. Figure 20 shows that including a transfer experiment in the flow cell procedure, only
minor changes occur in the current/potential diagram. The
areas under the oxidation peaks are the same within the experimental error (±5%).

Electrodes of 2×2 cm^2 geometric area were used in the experiments for Figs. 20 and 21. For the platinum electrode, a
pretreatment by fast triangular potential scans (200–300 V/sec
between 0.05 and 1.5 V RHE) followed by heating up to 900 K
in a 3×10^{-6} mbar O_2 atmosphere was applied. Electrodes

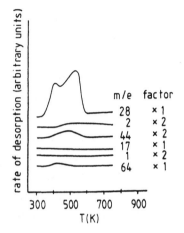
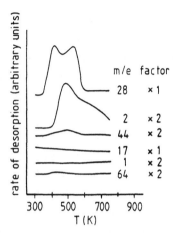

FIG. 21. ECTDMS of CO adsorbate (left side) and methanol adsorbate (right side) on platinum. Adsorption out of a 0.05 M H_2SO_4 solution at 455 mV. Desorption diagrams for m/e = 64 (SO_4), m/e = 44 (CO_2), m/e = 28 (CO), m/e = 17 (H_2O), and m/e = 2 (H_2); temperature scan rate 5 K/sec.

pretreated in this way can be immersed with a thin liquid film, easily removed by pumping in the lock chamber. The heat treatment reduces drastically the contamination of the platinum electrode by carbon. The roughness factor is usually in the order of three.

As an example, the direct proof of hydrogen in the methanol adsorbate at platinum in acid solution follows: First, in a blank experiment the platinum electrode was taken out of a H_2SO_4 solution (adsorption potential +455 mV) and immediately transferred to the UHV analysis position. During the thermal desorption procedure no signals were observed for masses 1 and 17 (H_2O), 44 (CO_2), 2 (H_2), and 28 (CO). After adsorbing carbon monoxide out of a H_2SO_4 solution (using again an adsorption potential of 455 mV), a pronounced double peak for the CO adsorbate was obtained while raising the temperature up to 700 K (see Fig. 21, left panel). Repeating the adsorption experiment in using a 5×10^{-2} M CH_3OH + 0.05 M H_2SO_4 solution and an adsorption potential of 455 mV, the TDMS signals of Fig. 21 (right panel) were obtained. In addition to the diagram for adsorbed CO, a strong desorption peak for m/e = 2 is recorded. Note that the H_2 signal is amplified by a factor of 2 compared to the CO signal.

The experiment of Fig. 21 clearly shows the existence of hydrogen in the methanol adsorbate under the described adsorption conditions. A quantitative calculation of the ratio $CO:H_2$ leads to 1:0.42. Therefore, it can be concluded that the hydrogen desorbed is due to the decomposition of a (C,O,H)-containing adsorbate. Obviously the methanol adsorbate on platinum is a mixture of CO and COH. Further investigations have shown that the ratio of CO:COH strongly depends on the methanol concentration in solution.

The ex situ technique described above offers new insight into the structure of the electrolytic double layer. At adsorption potentials in the hydrogen and oxygen region of a platinum electrode, one observes water molecules and anions (e.g., SO_4^{2-}, m/e = 64) in large amounts via the TDMS analysis.

REFERENCES

1. S. Bruckenstein and R. R. Gadde, J. Am. Chem. Soc. 93: 793 (1971).

2. M. Petek, S. Bruckenstein, B. Feinberg, and R. Adams, J. Electroanal. Chem. 42:397 (1973).

3. M. Petek and S. Bruckenstein, J. Electroanal. Chem. 47: 329 (1973).

4. S. Bruckenstein and J. Comeau, Discuss. Faraday Soc. 56:285 (1974).

5. R. R. Gadde and S. Bruckenstein, J. Electroanal. Chem. 50:163 (1974).

6. L. Grambow and S. Bruckenstein, Electrochim. Acta 22:377 (1977).

7. G. Kreysa and G. Breidenbach, Fresenius Z. Anal. Chem. 301:402 (1980).

8. O. Wolter, C. Giordano, J. Heitbaum, and W. Vielstich, Proc. Sym. Electrocatalysis, The Electrochem. Soc., Pennington, 1982, p. 235.

9. O. Wolter and J. Heitbaum, Ber. Bunsenges. Phys. Chem. 88:2 (1984).

10. S. Wasmus, E. Cattaneo, and W. Vielstich, Electrochim. Acta, 35:771 (1990).

11. K. Nishimura, R. Ohnishi, K. Kunimatsu, and M. Enyo, J. Electroanal. Chem. 258:367 (1987).

12. T. Iwasita, W. Vielstich, and E. Santos, J. Electroanal. Chem. 229:367 (1987).

13. W. Vielstich, Proc. Symposium Electrode Materials and Processes for Energy Conversion and Storage (S. Srinivasan, S. Wagner and H. Wroblowa, eds.), The Electrochemical Society Inc., Pennington, 1987, p. 394.

14. D. Tegtmeyer, A. Heindrichs, and J. Heitbaum, Ber. Bunsenges. Phys. Chem. 93:201 (1989).

15. Th. J. Brockman and L. B. Anderson, Anal. Chem. 56:207 (1984).

16. G. Hambitzer and J. Heitbaum, Anal. Chem. 58:750 (1983).

17. S. Wilhelm, W. Vielstich, H. W. Buschmann, and T. Iwasita, J. Electroanal. Chem. 229:377 (1987).

18. B. Bittins-Cattaneo, S. Wilhelm, E. Cattaneo, H. W. Buschmann, and W. Vielstich, Ber. Bunsenges. Phys. Chem. 92:1210 (1988).

19. J. Willsau and J. Heitbaum, J. Electroanal. Chem. 194:27 (1985).

20. A. E. Bolzán and T. Iwasita, Electrochim. Acta 33:109 (1988).

21. O. Wolter and J. Heitbaum, Ber. Bunsenges. Phys. Chem. 88:6 (1984).

22. B. Bittins-Cattaneo and T. Iwasita, J. Electroanal. Chem. 238:151 (1987).

23. J. Mrha, U. Vogel, S. Kreuels, and W. Vielstich, J. Power Sources 27:201 (1989).

24. G. Eggert and J. Heitbaum, Electrochim. Acta 31:1443 (1986).

25. P. Novák and W. Vielstich, J. Electrochem. Soc., in press.

26. P. Novák, P. A. Christensen, T. Iwasita, and W. Vielstich, J. Electroanal. Chem. 263:37 (1989).

27. D. W. DeWulf, T. Jin, and A. J. Bard, J. Electrochem. Soc. 136:1686 (1989).

28. J. J. Kim, D. P. Summers, and K. W. Frese Jr., J. Electroanal. Chem. 245:223 (1988).

29. J. Willsau, O. Wolter, and J. Heitbaum, J. Electroanal. Chem. 195:299 (1985).

30. M. Wohlfahrt-Mehrens and J. Heitbaum, J. Electroanal. Chem. 237:251 (1987).

31. C. R. Blackley and M. L. Vestal, Anal. Chem. 55:750 (1983).

32. K. J. Volk, R. A. Yost, and A. Brajter-Todt, Anal. Chem. 61:1709 (1989).

33. S. Wilhelm, H. W. Buschmann, and W. Vielstich, DECHEMA Monographie VCH Verlagsgesellschaft 112:113 (1988).

CARBON ELECTRODES: STRUCTURAL
EFFECTS ON ELECTRON TRANSFER KINETICS

Richard L. McCreery

The Ohio State University
Columbus, Ohio

I. INTRODUCTION AND SCOPE

Since the use of charcoal electrodes by Michael Faraday, carbon
has been a widely used and extensively studied electrode material for electrochemical applications. To provide a broad eco-
nomic perspective, about $35 billion of the 1985 U.S. gross na-
tional product resulted directly from electrochemistry [1], a significant fraction of which involved carbon electrodes. A recent
monograph [2] covers both fundamental and applied aspects of
carbon electrochemistry, with an emphasis on carbon black. A
symposium proceedings volume appeared in 1984 [3], and review
articles of interest to analytical chemists are in print [4-8].
These references establish the diversity of applications of carbon
materials in electrochemistry and provide a perspective for the
current chapter.

The suitability of an electrode material for analytical applications is dictated by electron transfer rate, background current,
mechanical properties, stability, etc. Analytically useful electrodes may differ substantially from those used in electrosynthesis or fuel cells, since the objectives are quite different. In
many practical cases, the choice of a carbon material for analytical applications is determined by convenience or empirical observations, without regard to carbon structure. Due mainly to
difficulties in preparing and characterizing well-defined carbon
surfaces, there has been limited progress in relating surface
structure to electroanalytical performance.

This chapter is divided into four major sections, with the
overall goal of describing a relationship between carbon electrode
structure and electroanalytical performance. Particular emphasis
will be placed on structural variables that affect electron transfer kinetics and voltammetric background current, since these
variables usually determine the analytical utility of an electrode.
In the first major section, the structure and bulk properties of

various sp^2 carbon materials will be described. In the second section, the carbon surface will be considered, with particular attention to surface structure and characterization. Third, the electrochemical consequences of bulk structure and surface preparation will be discussed, with particular emphasis on the relationship between surface structure and heterogeneous electron transfer rate. Finally, a model of the carbon electrode surface will be proposed which accounts for the experimental observations. Rather than attempt to comprehensively discuss carbon structure and properties, the discussion will be limited to those aspects of carbon which are known to affect electrode kinetics and background current. In order to keep the scope of the chapter at a manageable level, the discussion will be limited to experiments in aqueous media. With the exception of surface oxidation, chemical modifications of carbon electrodes will not be addressed. These constraints on scope force selectivity in the choice of literature citations and the exclusion of a large body of valuable published results. The omission of many excellent papers involving carbon electrodes permits a tighter focus on the effects of carbon structure on electron transfer kinetics.

II. BULK CARBON STRUCTURE AND PREPARATION

Nearly all of the many types of sp^2 hybridized carbon have been used in electrochemistry, including carbon black, glassy carbon, randomly oriented graphite, pyrolytic graphite, etc. Although the carbon−carbon bonding of all these materials is similar, the bulk properties of the materials vary greatly due to the size and orientation of graphitic crystallites. After a brief discussion of relevant structural variables in bulk sp^2 carbon, several prominent types of carbon will be described.

A. Structural Variables in sp^2 Carbon

1. Intraplanar C–C Bond Length

The C–C bond length varies from 1.39 Å in benzene to 1.421 Å
in graphite [9], and is essentially equal for all types of sp^2
carbon. Although one would predict some changes in the C–C
bond length near the edges of graphite planes, such variations
are difficult to measure and have not been determined. In the
context of electrochemical applications of sp^2 carbon, the bulk
C–C bond length is 1.42 Å.

2. Intraplanar Microcrystallite Size, L_a

As shown in Fig. 1, L_a is the mean size of the graphitic microcrystallite along the a-axis, which always lies in the plane of
the hexagonal lattice. L_a is determined from x-ray diffraction
line widths [10] and may vary from 10 Å up to about 10 μm
(10^5 Å) or more. Strictly speaking, L_a is the average coherence

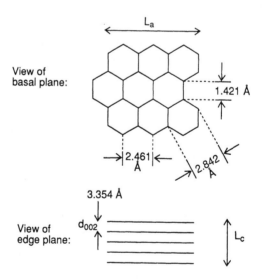

FIG. 1. Crystallographic dimensions of sp^2 carbon. L_a, L_c,
and d_{002} vary with carbon type.

length of the lattice giving rise to a particular diffraction peak, but in most discussions of carbon, the coherence length is assumed to equal the mean crystallite size. The Raman spectrum of carbon depends strongly on L_a, such that L_a may be determined spectroscopically in the range of ca. 25–1000 Å [11]. Since L_a can vary almost continuously from crystallites only slightly larger than benzene up to crystallites containing millions of aromatic rings, one should consider L_a to define a near continuum of molecules with varying properties. Thus, any sp^2 carbon sample contains a variety of molecules with varying L_a, and there is no reason to expect carbon samples with different L_a's to behave identically.

3. Interplanar Microcrystallite Size, L_c

A perfect single crystal of hexagonal graphite would have large L_a, and the graphite planes would stack in an ABAB. . . sequence. The distance along the axis perpendicular to the graphite planes over which this sequence is realized is defined as L_c, with the "c"-axis always perpendicular to the graphite planes. An alternative ABCABC . . . stacking occurs in rhombohedral graphite, but this material is rare (a few %) in natural and synthetic graphite [12] and will be ignored here. In severely disordered carbons (e.g., carbon black), L_c may cover only a few graphitic layers and be less than 10 Å, while for natural graphite single crystals L_c may exceed 10 µm. As with L_a, L_c can have pronounced effects on the properties of the carbon. A relevant example is provided by the "graphitization" process. When disordered (i.e., small L_c) carbon is heated to above 2000°C, L_c increases, and the material is said to "graphitize." As the sp^2 carbon planes stack parallel to each other, the material becomes softer and L_c increases. Since interplanar bonding is through weak van der Waals' forces, the planes can move easily with respect to each other, and graphite is a useful

lubricant. The difference between carbon black and graphite
is primarily the larger L_C caused by heating.

4. Intraplanar Spacing, d_{002}

The label d_{002} is derived from the x-ray diffraction designation
of the reflection corresponding to the interplanar spacing. The
interplanar spacing is 3.354 Å for single crystal graphite [12]
but can range up to 3.6 Å or more for less ordered sp^2 carbon.
When graphite is intercalated, d_{002} can be much larger, up to
ca. 10 Å [13]. d_{002} is most commonly determined with x-ray
powder diffraction, although occasionally neutron scattering is
employed. Most synthetic graphite is said to be turbostratic,
with a slightly larger mean value of d_{002}. The ABAB stacking
sequence may be observed for short L_C distances, but frequently
the layers are rotationally disordered. The planes are still parallel and d_{002} ranges from 3.354 to 3.359 Å [9]. Most literature
references regard L_C as the distance over which parallel stacking
occurs, ignoring the distinction between turbostratic and hexagonal stacking.

When one is dealing with sp^2 carbon with crystallites containing more than 50 rings or so, the bulk intraplanar C–C
distance can be assumed to be 1.42 Å. That leaves three independent variables which describe the structure of the material: L_a, L_c, and d_{002}. It should be noted, however, that
each of these variables has a range of values and that knowledge
of their means does not uniquely describe a given carbon material. Nevertheless, knowledge of L_a, L_c, and d_{002} is usually
sufficient to predict many physical and electrochemical properties of the materials. A corollary to this statement is the observation that sp^2 carbon material with differing L_a, L_c, or
d_{002} may have quite different physical and electrochemical behavior, and one should be careful to determine or control these
variables when using carbon as an electrode.

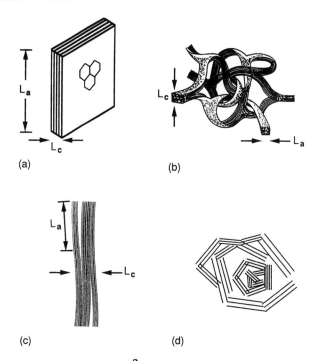

FIG. 2. Four types of sp^2 carbon. Drawings are not to scale.
(a) HOPG; (b) glassy carbon; (c) carbon fiber (edge view);
(d) carbon black.

Figure 2 shows several distinct carbon materials, and Fig. 3
classifies carbon materials according to L_a and L_c.

B. Bulk Characterization of sp^2 Carbon

Before discussing different types of sp^2 carbon, two characterization techniques commonly used to determine bulk carbon
structure deserve special note. The most commonly used method for determining L_a, L_c, and d_{002} is x-ray powder diffraction
[10,14]. The subscript 002 refers to the appropriate reciprocal
lattice vector of the graphite crystal. d_{002} is determined

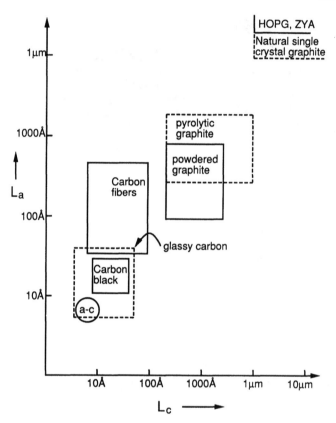

FIG. 3.   Approximate ranges of L_a and L_c for various sp^2
carbon materials.   Since there is large variation of L_a and L_c
with sample history, the ranges shown should be considered
representative but approximate.   a-c is amorphous carbon; ZYA
is the highest available commercial grade of highly ordered pyrolytic graphite.

directly from the Bragg equation applied to powder diffraction from a graphitic sample. By analogy to Fourier transforms, one can appreciate that a periodic lattice of finite size will lead to peak broadening in the diffraction pattern. Thus the linewidth (in radians) of the 002 reflection is related to L_c by Eq. (1).

$$L_c = \frac{0.9\,\lambda}{\beta\cos\Theta} \tag{1}$$

where β and Θ are the linewidth and angle of the 002 reflection, respectively, and λ is the x-ray wavelength. L_a may be determined similarly from an hk0 reflection. L_a and L_c determined from x-ray linewidth are mean values and are subject to some misinterpretation [14]. For example, carbon fibers may have crystallites which have a large L_a along the fiber axis, but a small L_a perpendicular to the axis, i.e., the fiber may be composed of long, ribbonlike crystallites. Clearly a mean value of L_a does not precisely describe the microstructure of such a fiber. Nevertheless, L_a and L_c are commonly used to characterize carbon materials and are often stated directly in terms of the diffraction peak linewidth as the "mosaic spread" of the hk0 peak in degrees. A small mosaic spread ($>1°$) implies a highly ordered material with large L_a. For example, the highest commercial grade of highly ordered pyrolytic graphite (HOPG) has a mosaic spread of $<0.4°$ and an L_a of >1 μm.

Raman spectroscopy is a more recent addition to the battery of techniques used to characterize carbon, but provides quite useful information. Carbon has a weak IR adsorption spectrum because most lattice vibrations yield weak dipole moment changes. Since Raman selection rules are based on polarizability rather than dipole moment, graphitic carbon is a relatively strong Raman scatterer. A typical arrangement for Raman spectroscopy of strongly absorbing carbon samples is shown in Fig. 4.

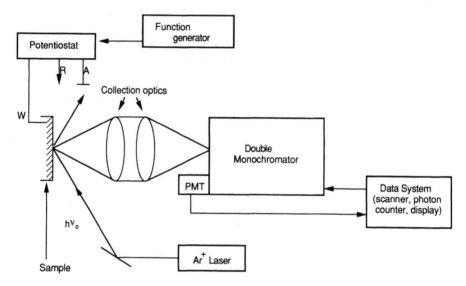

FIG. 4.   Block diagram of instrumentation for in situ Raman spectroscopy of solid electrodes.   Sample may be in air or electrolyte, with potential control if desired.

The laser beam penetration depth for normal incidence is given by Eq. (2).

$$I = I_0 \exp(-4\pi k z/\lambda) \tag{2}$$

where

I = laser intensity at some depth z
I_0 = laser intensity at surface
λ = laser wavelength
k = absorption coefficient of carbon at wavelength λ

The Raman scattered light must also pass through the absorbing carbon, so a similar equation applies to the escape of Raman light.   Assuming k and λ do not differ greatly for the scattered vs. incident light, the Raman signal, S_R, will be given by Eq. (3) [15,16] for a sample of thickness z.

$$S_R = S_{max} \left(1 - \exp \frac{-8\pi kz}{\lambda} \right) \tag{3}$$

where S_{max} is the signal observed at large z. Values of k have been reported as 1.52 for HOPG at 515 nm [16] and 0.691 for GC at 500 nm [17]. Equation (3) can be used to determine that 63% of the maximum Raman signal (i.e., $8\pi kz/\lambda = 1$) will be obtained at the depths (Z_{63}) shown in Table 1. Since k varies slowly with wavelength in the visible/NIR range [17], sampling depth increases with wavelength. As a rule of thumb, the Raman sampling depth is about 300 Å when visible light is used. This value is much smaller than many optical techniques applied to solids because of the strong optical absorption of carbon. The significant efforts of several research groups on the Raman spectroscopy of carbon will be summarized here, insofar as they relate to electrode materials. Published Raman spectra of carbon are most commonly obtained in air with visible (488 or 515 nm) laser light.

TABLE 1

Effective Sampling Depths for Raman Spectroscopy of Carbon

	λ(nm)[a]	k	Z_{63}(Å)[b]
HOPG	515	1.52	135
GC	416	0.68	243
GC	500	0.69	284
GC	515	0.70	293
GC	625	0.74	336
GC	1250	1.04	478

[a]Assumes normal incidence of laser beam.
[b]Sample thickness producing (1 − 1/e) of the maximum Raman signal.

Based on the D_{6h}^4 symmetry of an infinite graphite crystal, the following vibrational modes are possible [11]:

$$2B_{2g} + 2E_{2g} + A_{2u} + E_{1u} \tag{4}$$

These vibrational modes are shown schematically in Fig. 5, and only the two E_{2g} modes are predicted to be Raman active. The Raman spectrum of HOPG shown in Fig. 6a exhibits a well defined peak at 1582 cm^{-1}, which has been assigned to one of the

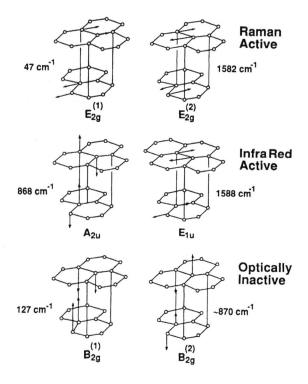

FIG. 5. Vibrational modes for a graphite crystal predicted for the D_{6h}^4 space group. Frequencies were determined from Raman and neutron scattering experiments. The 870 cm^{-1} mode frequency is predicted but has not been observed experimentally. (Adapted from Ref. 23.)

E_{2g} modes [11,18–22]. The other E_{2g} mode is at very low frequency (42 cm^{-1}) and is difficult to observe [23]. The peak(s)
above 2600 cm^{-1} are assigned to overtones and combinations,
and their assignments are somewhat controversial [19,24–27].
Most authors agree that the 1582 cm^{-1} Raman peak is first order
scattering of E_{2g} symmetry for a material with large L_a and L_c.
When L_a is decreased (as in glassy carbon, Fig. 6b) a new feature at 1360 cm^{-1} is observed. As shown in Fig. 7, the ratio
of the 1360 cm^{-1} to 1582 cm^{-1} integrated peak intensities is directly proportional to $1/L_a$, determined crystallographically [11].
While there is general agreement on the increase of the 1360 cm^{-1}
intensity with decreasing L_a, the mechanism is not obvious. The
1360 cm^{-1} band has been assigned to an A_{1g} mode, which is not
predicted to be observed in an infinite graphite sheet [11]. A
related explanation is based on symmetry breaking, which occurs
at the edges of graphite sheets. The edges lead to breakdown
of the selection rules for optical excitation of phonons in the
graphite lattice [13,18,19]. More phonons can result in Raman
scattering in the disordered material, including those at ~1360
cm^{-1}. The more edge present (as in small microcrystallites),
the more 1360 cm^{-1} intensity, thus the correlation of the I_{1360}/
I_{1582} intensity ratio with $1/L_a$. Based on polarization sensitive
Raman spectra of HOPG edge plane, Katagiri et al. [28] concluded that the edges of large crystallites produce the 1360 cm^{-1}
peak, confirming the correlation of this band with edge plane
carbon rather than small microcrystallites, per se. Although
there remains some controversy on the origin of the 1360 cm^{-1}
band, there is agreement on the useful correlation of 1360 cm^{-1}
band intensity with graphitic edge plane density.

The E_{2g} peak at 1582 cm^{-1} broadens and shifts to higher
frequency with decreasing L_a and L_c. This shift can be correlated with microstructure and heat treatment history to some extent [26], but specific conclusions are somewhat elusive. One

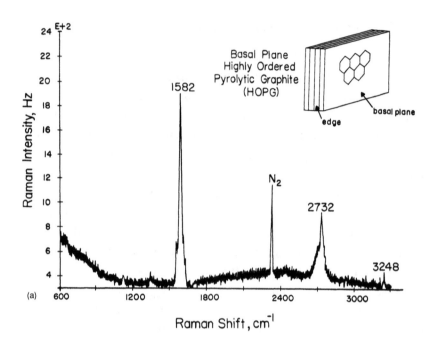

(a)

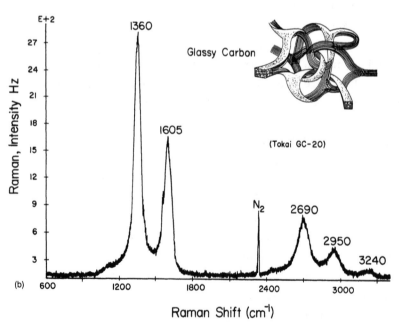

(b)

FIG. 6.

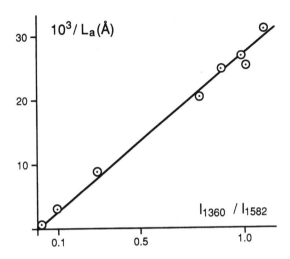

FIG. 7. Relationship between L_a and the ratio of the integrated Raman intensities for the 1360 cm^{-1} and 1582 cm^{-1} bands. (Adapted from Ref. 11.)

effect of structure on the E_{2g} band is robust, however, and can be useful for characterizing carbon samples. When the graphitic planes are separated by a layer of intercalant, the E_{2g} frequency shifts to ca. 1620 cm^{-1}. Dresselhaus et al. [29] attribute this shift to "boundary layer" graphite planes which are not adjacent to other graphite layers. For example, a stage 4 intercalation compound has four graphite layers between each layer of intercalant, so half the graphite layers are next to graphite only, and half are next to intercalant. The Raman spectrum (Fig. 8) shows equal intensities of 1582 and 1620 cm^{-1}

FIG. 6. Raman spectra obtained in air for HOPG and GC with a 515 nm laser in both cases. a) Raman spectra obtained in air for HOPG with a 515 nm laser; b) Raman spectra obtained in air for GC with a 515 nm laser.

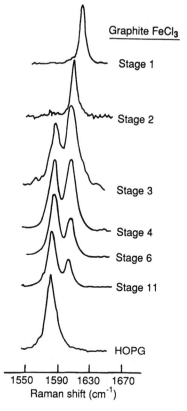

FIG. 8.   Effect of $FeCl_3$ intercalation on E_{2g} mode of graphite.
(Adapted from Ref. 29.)

peaks for the stage 4 intercalation compound.   Although the
magnitude of the frequency shift is slightly dependent on the
identity of the intercalant, it is always a shift to higher frequency [29,30].   Combined with the observation that the E_{2g}
mode broadens and shifts to higher frequency as d_{002} increases,
one concludes that the 1620 peak is due to exfoliation (or delamination) of the graphite layers [31].   Thus the E_{2g} frequency
and linewidth provide a useful semiquantitative marker for inter-
layer graphite spacing.

In summary, the $I_{1360/1582}$ ratio is correlated with edge plane density, and the linewidth and position of the 1582 cm^{-1} band indicate stacking disorder. In general, $I_{1360/1582}$ is a marker for L_a, while the position and linewidth of the E_{2g} mode are related to L_c and d_{002}. These observations are of significant value in characterizing carbon electrodes, particularly since Raman spectroscopy can be used in situ during electrochemical processes.

C.   Preparation and Properties of Bulk Carbon Materials

In this section, several types of sp^2 carbon useful for electrochemistry will be described. The materials will be listed according to the degree of microstructural disorder, with the most ordered materials first. If the reader desires more breadth and detail than presented here, excellent reviews are available on HOPG [32], electronic properties of carbon [33], carbon fibers [34], glassy carbon [35], carbon black [2], and amorphous carbon [9].

1.   Single Crystal Graphite

The only current sources of graphite crystals are in natural geologic formations, and natural crystals are primarily hexagonal, with the ABAB . . . stacking pattern [32]. As expected from the structure, many properties of graphite crystals are extremely anisotropic, with the thermal and electrical conductivities being much higher along the a-axis than the c-axis. Since electron mobility in the extended π system in the graphite plane is high, the a-axis electrical conductivity is high, about 4% that of copper [32]. The values of the a- and c-axis resistivities for natural graphite are somewhat variable due to impurities and defects in the crystal. The c-axis resistivity (ρ_c) is particularly prone to variability, since a defect can provide a low resistance path in what is usually a relatively high resistance material. The resistances of two natural graphite crystals are listed in Table 2. Keep in mind that individual samples of natural graphite can vary significantly in ρ_c and ρ_a, sometimes by orders of magnitude. The variability of properties and supply of natural graphite crystals stimulated the development of synthetic materials based on pyrolytic graphite.

2. Pyrolytic Graphite (PG) and Highly Ordered Pyrolytic Graphite (HOPG)

Pyrolytic graphite is a somewhat generic term applied to carbon materials prepared by high temperature decomposition of gaseous hydrocarbons onto a hot surface. At deposition temperatures above 1200°C, significant ordering of the a-axis parallel to the hot substrate occurs, and the material is known as pyrolytic graphite. The temperature is often increased during or after deposition to as high as 3800°C to increase L_a, L_c, and the degree of ordering. If the deposition temperature is too low, a disordered pyrolytic carbon will result. PG has distinctly anisotropic properties due to orientation of graphite planes, but the anisotropy is not as pronounced as HOPG.

HOPG, also known as pressure annealed pyrolytic graphite, is made by annealing PG at high temperature under high compression force along the c-axis. The process was developed by Union Carbide, and HOPG is sold primarily for use as x-ray

TABLE 2

Resistivities of Ordered Graphite Materials

	$\rho_c(\Omega - cm)$	$\rho_a(\Omega - cm)$	ρ_c/ρ_a
Single crystal, Ticonderoga	3.0×10^{-3}	3.8×10^{-5}	80
Single crystal, Kish	2.5×10^{-1}	1.0×10^{-4}	2500
HOPG, annealed at 3800°	1.7×10^{-1}	4.1×10^{-5}	~3800

Source: From Ref. 32.

diffraction gratings. Several grades are available based upon the linewidth of the d_{002} x-ray diffraction peak, with the highest grade (ZYA) having a mosaic spread of $0.4 \pm 0.1°$ (1.54 Å light). HOPG is "turbostratic," meaning the graphite layers are rotationally disordered and the ABAB pattern present in hexagonal graphite is observed over only short distances along the c-axis. As mentioned earlier, most authors use L_c to refer to the parallel planes of graphite, without regard for the presence of the ABAB . . . pattern [36]. In this chapter, the conventional reference to L_c as the coherence length for parallel stacking will be used. Typical L_a and L_c values for ZYA HOPG are 2 µm and 10 µm, respectively. An exposed hexagonal surface that is perpendicular to the c-axis is known as a basal plane, while a cut surface that is parallel to the c-axis is an edge plane.

As shown in Table 3, most properties of HOPG are quite anisotropic. Although these properties vary with the quality of the HOPG sample (i.e., the number of defects), they are much less variable than those of natural crystals. The higher resistivity ratio of HOPG compared to natural crystals (see Table 2) implies that HOPG more closely approximates an ideal crystal. It is clear from Table 3 that the a-axis conducts heat and electrons much better than the c-axis, is much more resistant to mechanical stress, and has a much smaller thermal expansion coefficient. Since the c-axis properties, particularly resistivity, are affected by defects, the anisotropy of a perfect graphite crystal may be higher than indicated in Table 3. The basal plane surface of HOPG is black but shiny to the eye. As demonstrated by numerous experiments with scanning tunneling microscopy [39,40], the basal surface is nearly atomically smooth. In contrast, the edge plane is rough and dull, and it is difficult to prepare in any manner which is atomically well defined. The ease of observation of carbon atoms on HOPG in air with tunneling

TABLE 3

Anisotropic Properties of HOPG

	c-axis	a/axis	c/a ratio	Ref.
Resistivity (Ω – cm)	1.7×10^{-1}	4.1×10^{-5}	~3800	32
Thermal conductivity (W/cm-sec-°K)	0.10	24	0.004	37
Thermal expansion (10^{-6}/°C)	27	$-1/2$[a]	23	37
Young's modulus (MPa $\times 10^{-5}$)	0.3	10.3	0.03	38

[a]Temperature dependent; becomes positive above 1000°C.

microscopy implies that the basal plane of HOPG is remarkably inert toward adsorption of impurities, in contrast to metals.

The electronic band structure of HOPG is unusual, and has received significant attention in the solid state physics community [9,32]. HOPG is a semimetal, with a small overlap (0.04 eV) of the conduction and valence bands. This fact leads to some anomalous magnetic and electrical effects which are beyond the scope to this chapter [32]. However, one unusual property of HOPG is the low double layer capacitance observed on the basal plane [4,41–43].

3. Randomly Oriented Graphite

The most common forms of graphite are powders and molded shapes, with random orientation of the a- and c-axes of microcrystallites. The term "randomly oriented graphite" will be used here to mean graphitized carbon with randomly oriented microcrystallites, and should be distinguished from "amorphous carbon," in which very little ordering of carbon planes has occurred. The microcrystallites of randomly oriented graphite have the expected 3.35 Å spacing, with small L_a and L_c. Powdered graphite can be made by graphitizing carbon black at high temperature (>2500°C), and a very wide range of powders are available commercially. Molded graphite parts (rods, disks, cups, etc.) are made by heat treating carbon-rich precursors made from petroleum pitch [2]. The coke and pitch fractions of petroleum are low volatility mixtures of large, heavy polynuclear aromatic hydrocarbons. Upon heating to high temperatures (2500–3000°C), the large hydrocarbons fuse to form larger sp^2 carbon planes, then graphitize, thereby increasing both L_a and L_c. After shaping a mixture of coke and pitch (the two materials differ in average molecular weight, with pitch being higher) into the desired form, the piece is graphitized to make the finished graphite part. Since nothing about the procedure leads to a preferential orientation, the part consists of randomly oriented microcrystallites, unlike PG or HOPG. As shown in Table 4, randomly oriented graphite has lower density and higher resistivity than HOPG. Most graphite is porous and permeable to gases and in some cases liquids. A typical solid graphite part such as a rod may be 15–30% porous, leading to the lower apparent density. Variations in the manufacturing process will lead to variations in density, L_a, and L_c, yielding a wide range of graphitic materials with verying properties. Since there are many manufacturers offering many grades of graphite, one must use care in choosing one for an electrochemical application. In some cases, the types of graphite will not affect the outcome of an experiment, but it is nevertheless prudent to control the variables related to the graphite when possible.

4. Glassy Carbon

Glassy carbon (GC) is also known as vitreous carbon, and may be purchased as disks, rods, or as highly porous reticulated vitreous carbon (RVC). RVC is microstructurally very similar to GC but is heat treated in a process which leaves large voids in a solid network that is as much as 95% air. Unlike graphite, GC is hard and impermeable to gases or liquids, and has slightly lower electrical and thermal conductivities. GC does not graphitize, even at a temperature of 3000°C.

The structural differences between GC and graphite are best appreciated by considering the synthetic route to GC. While graphite is prepared from hydrocarbons or petroleum fractions, GC is made from polymeric resins such as polyacrylonitrile (PAN) or phenol/formaldehyde polymers. The polymer is heat treated at 1000–3000°C, often under pressure, and any H, N, or O atoms are released, leaving an extensively conjugated sp^2 carbon structure (Fig. 9). The original polymeric backbone stays largely intact, preventing the formation of extended graphitic domains. The result is a complex structure of interwoven

TABLE 4

Properties of Various Carbon Materials

	Apparent density (g/cc)	d_{002} (Å)	ρ (Ω-cm)	L_a (Å)	L_c (Å)
HOPG, a-axis	2.26	3.354	4×10^{-5}	>10,000	>100,000
HOPG, c-axis			0.17		
Pyrolytic graphite	2.18	3.34		1000 (typ)[a]	1000 (typ)
Randomly oriented graphite (Ultracarbon UF-4s grade)	1.8	3.35	1×10^{-3}	300 (typ)	500 (typ)
Tokai GC-10[b]	1.5	3.49	4.5×10^{-3}	20 (typ)	~10
Tokai GC-20[b]	1.5	3.48	4.2×10^{-3}	25 (typ)	12
Tokai GC-30[b]	1.5	3.41	3.7×10^{-3}	55	70
Carbon fiber (typ)	1.8	3.4	$(5-20) \times 10^{-4}$	>100	40
Carbon black (Spheron 6)	$(1.3-2.0)^{c}$	3.55	0.05	20	13
Evaporated a-C	~2.0	>3.4	~10^3	~10	~10
a-C:H	1.4 – 1.8		$10^7 - 10^{16}$		

[a] Entries marked "typ" may vary significantly with sample or preparation procedures.
[b] Number refers to heat treatment temperatures, e.g., GC-20 was treated at 2000°C.
[c] Depends on technique used to measure density [2].
Source: From Ref. 46.

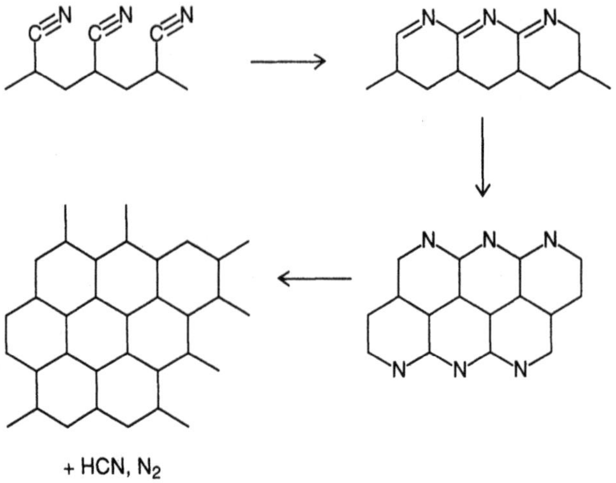

FIG. 9. Highly simplified reaction sequence leading from poly-acrylonitrile to sp^2 carbon. Many steps involved are not shown. (Adapted from Ref. 34.)

graphitic ribbons, with L_a typically 50 Å and L_c 15 Å. The popular model for GC proposed by Jenkins and Kawamura [35] is shown in Fig. 2. The ordered layers occurring in graphitic domains have an average d_{002} greater than 3.354 Å, but are extensive enough to yield reasonable electrical conductivity, about one-fourth that of randomly oriented graphite. The in-terwoven sp^2 carbon ribbon results in mechanical hardness, since distortion of the material requires bond breaking. The density of ca. 1.5 for GC is much lower than HOPG [2,35], indicating ca. 35% void space. Since GC is impermeable to gases, the voids must be small and unconnected.

Many properties of GC are dependent on the heat treatment temperature (HTT), and this temperature is usually specified when a GC sample is sold. A clear demonstration that HTT is affecting GC microstructure is provided by the Raman data of Fig. 10. In addition to minor effects on the shape and intensities

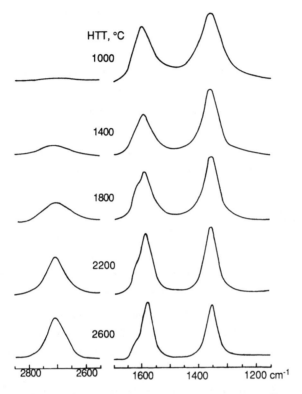

FIG. 10.   Effect of heat treatment temperature on Raman spectrum of glassy carbon.   (Adapted from Ref. 25.)

of the 1360 cm^{-1} and E$_{2g}$ modes, HTT drastically affects the intensity of the 2730 cm^{-1} band.   The structural implications of these spectral changes are not completely clear, but they are usually interpreted in terms of greater ordering of the GC at higher temperature [25,26].   The 2730 cm^{-1} band is destroyed by ion bombardment and can be restored by reheating.   Lespade et al. [26] have attributed these spectral changes to localized graphitization, with increases in both L$_a$ and L$_c$ occurring at higher HTT.   The importance of these structural effects to electrochemical applications is currently not known, but is clear that thermal history is of major importance to GC structure. A user of GC is prudent to be aware of HTT as a possible variable in electrode performance.

In addition to variations in L_a, L_c, and d_{002} caused by variable HTT, samples of GC can also vary in the extent of macroscopic defects present. A common occurrence is pits of a few μm diameter, which apparently result from gas bubbles formed during heat treatment. In addition, soft regions of what appears to be graphite can occasionally be observed during polishing of a GC surface.

5.   Carbon Fibers

The high tensile strength and light weight of fine carbon fila-ments have resulted in extensive development of preparation techniques and numerous reports on carbon fiber properties [34]. Carbon fibers range in diameter from 0.1 to 100 μm, with the common range being 5–25 μm. The many procedures for making carbon fibers can be classified into two types: heat treatment of polymeric precursors and catalytic chemical vapor deposition (CCVD). Because of applications requiring high tensile strength, the preparation of carbon fibers is usually designed to align the a-axis of the final graphitic lattice along the axis of the fiber, to take advantage of the high modulus along the a-axis.

Heat treatment procedures are similar to the methods used to prepare GC, except for important mechanical manipulations. For example, PAN or pitch is heated to about 300°C and the molten polymer is spun or extruded to form fibers. If these fibers are kept under tension during further heat treatment, the a-axis tends to align with the fiber axis. Heat treatment to 3000°C graphitizes the carbon, after driving off all non-carbon elements. Higher heat treatment temperatures generally result in better orientation of the a-axis and higher Young's modulus [44]. Weight loss during fiber manufacture depends

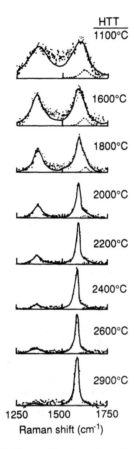

FIG. 11. Effect of HTT on first order Raman spectrum of carbon fiber made from benzene. (Adapted from Ref. 34, p. 99.)

on the precursor and can range from 75% (rayon) to 20% (pitch). The effect of HTT on Raman spectra of fibers is shown in Fig. 11. Note that the 1360 cm^{-1} peak intensity decreases as the fiber becomes more ordered, and the spectra of fibers prepared at high temperature are similar to HOPG.

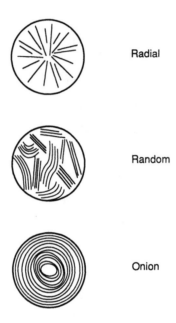

Radial

Random

Onion

FIG. 12.   Schematic representations of carbon fiber cross sections. Lines denote edges of graphite planes in all cases.

The fiber microstructure can vary significantly with the manufacturing procedure and several common fiber cross sections are shown in Fig. 12.   An additional variable not evident in Fig. 12 is the "micro-orientation angle" between the fiber axis and the a-axis.   All of these variables significantly affect fiber properties and can vary greatly with preparation procedure. Comparing Figs. 12 and 2, a key difference between GC and carbon fibers is evident.   GC is isotropic on a distance scale greater than about 100 Å, since there is no preferred orientation introduced by the manufacturing process.   Fibers are designed to have a preferred orientation, however, and possess additional structural variables related to orientation of the graphite planes.   The ordering in fibers along the fiber axis can extend at least several hundred Å.

The CCVD process for making carbon fibers is based on catalytic dehydrogenation of hydrocarbons (e.g., benzene) on small particles of Fe, Ni, Co, etc. [34]. When a small metal particle is heated in the presence of benzene vapor, for example, the fiber grows and elevates the particle, as shown in Fig. 13. The resulting fiber is often tubular, with an onion structure similar to Fig. 12. The growth mechanism is currently under investigation, but it appears to involve diffusion of carbon atoms on the surface or through the bulk of the metal particle after decomposition of the reagent gas. After diffusion, the carbon atoms precipitate to form graphite on the lower cooler

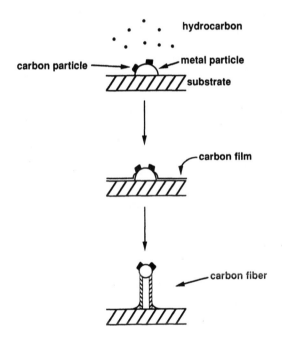

FIG. 13. Putative mechanism for catalytic chemical vapor deposition of hydrocarbons to form carbon fiber. Transport of carbon particle on metal is by surface or volume diffusion. (Adapted from Ref. 34.)

side of the metal particle. CCVD fibers are usually stronger
but shorter than those made from polymeric precursors, with
greater long-range order.

6. Carbon Black

Although its use in analytical electrochemistry is fairly uncommon, carbon black accounts for a significant fraction of the U.S.
production of carbon, and most of that is used in the rubber
industry. There exist six major synthetic routes to carbon
black, each of which can produce a variety of properties in the
final product, depending on conditions [45]. In all cases a
hydrocarbon is burned or thermally decomposed in the gas
phase, and carbon black deposits on the vessel walls. Carbon
black particles range in size from 300 to 5000 Å, but these
particles consist of smaller microcrystallites with L_a and L_c
equalling 10–50 Å and a d_{002} of about 3.5–3.6 Å. The surface
area of carbon black particles is very high, ranging up to
5000 m^2/g.

The structure of carbon black is very disordered compared
to graphite, with microcrystallites having only a few parallel
layers. All properties and in some cases the structure of carbon black depend on the conditions of manufacture. Resistance
is typically an order of magnitude or more higher than GC and
at least two ordrs of magnitude higher than HOPG (see Table
4). The generic structure for carbon black shown in Fig. 2
contains a disordered collection of very small microcrystallites,
leading to the low apparent density and conductivity. The
reader should consult Kinoshita's monograph for further information and references on carbon black [2].

7. Carbon Films

Carbon films have been studied for decades, but their use in
electrochemistry is only occasional. The recent surge of research into diamond film formation has led to further attempts

to control and characterize carbon film formation. A comprehensive review of carbon film electronic properties has appeared [9].

As noted earlier, thermal decomposition of hydrocarbons under conditions which lead to graphitization results in pyrolytic graphite. If the carbon is deposited from the gas phase after electron beam evaporation or sputtering onto a surface held below the graphitization temperature, a disordered film is formed which is usually called amorphous carbon (a-C). a-C is sometimes referred to as "hard carbon," or, unfortunately, "diamondlike" carbon. Given the recent interest in true diamond films, hard carbon is the preferred term for a-C. a-C is a semiconductor, has a lower density than graphite, but is not as hard as GC (see Table 4). The distinctive property of a-C compared to PG is the presence of 1–10% sp^3 bonded carbon. The sp^3 character results in the observed decrease in conductivity and increased hardness. Carbon films with even greater hardness are made by plasma deposition from a CH_4 or CH_4/H_2 atmosphere, and the resulting films have significant hydrogen content. Hydrogenated amorphous carbon (a-C:H) has lower conductivity, greater hardness, and lower density than a-C. Its low conductivity is caused by a significantly higher percentage of sp^3 sites (ca. 30%) compared to a-C.

It is clear that vapor deposited carbon films have properties which range from soft, conductive HOPG to hard, semiconducting a-C:H, depending on deposition technique and conditions. Ungraphitized carbon films have the greatest disorder and least well-defined structure of the carbon materials discussed here, and the properties of such films will be the most difficult to control. In order to summarize the description of carbon properties, the reader is directed to Figs. 2 and 14 and Table 4. It is clear that the electrochemist has a variety of materials to choose from, and that he should be aware of the variety of

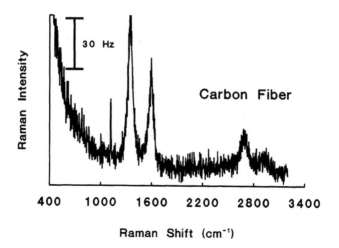

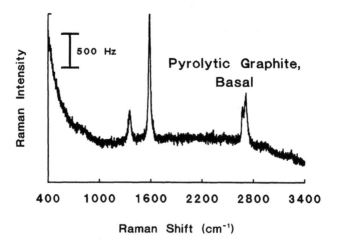

FIG. 14.   Raman spectra (515 nm) of several carbon materials.
Carbon fiber was heat treated ex-pitch type; graphite rod was
spectroscopic graphite (Zeebac #HP-242, grade Al); PG is from
Pfizer, basal plane; GC samples were from Tokai, polished with
Al_2O_3 on polishing cloth.   Sloping baseline at low Raman shift
is due mainly to elastic scattering.

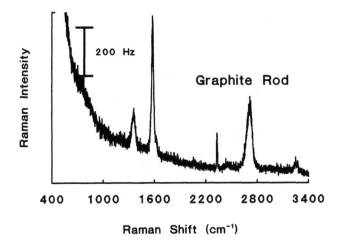

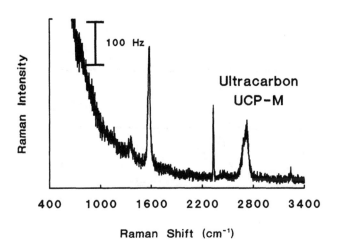

FIG. 14 (Continued).

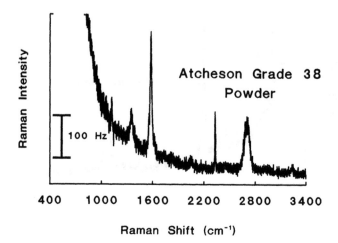

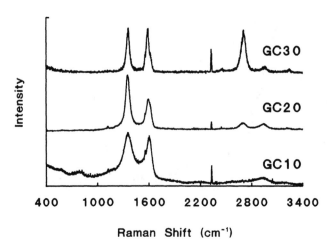

FIG. 14 (Continued).

properties. Figure 14 presents a collection of Raman spectra of carbon materials of electrochemical interest. Based on the 1360/1580 Raman intensity ratio, one concludes that L_a may vary from <30 Å (polished GC-20) to >1000 Å (HOPG). There is substantial variation of L_a in different graphite samples, and carbon fibers prepared differently may have very different L_a values (see Figs. 11,14). As we will see, these microstructural variations can substantially affect electrochemical behavior.

III. STRUCTURE OF CARBON ELECTRODE SURFACES

Carbon electrodes differ significantly from those made of metal both because of carbon's anisotropic structure and because of major differences in behavior toward oxidation. As discussed in the previous section, a variety of bulk carbon structures exist, each of which may yield one or more different surface structures when prepared for use as an electrode. The formation of surface oxides, CO, or CO_2 on carbon electrodes is usually chemically irreversible, and a variety of surface oxides are usually present on carbon. Despite the importance of surface chemistry to any electrochemical experiment, surface structural information is usually unavailable for carbon electrodes, and the majority of electroanalytical experiments are performed in the absence of a well-defined surface. However, a great deal of important information about the relationship between surface prepartion and electrochemical behavior has been obtained, and useful inferences about the effect of carbon surface structure on electron transfer kinetics are available. This section will discuss surface characterization with emphasis on current knowledge about surface structure. After discussing the surface variables which may affect electroanalytical performance, the effects of surface preparation methods on those variables will be addressed.

A.   Principal Surface Variables

The long-range objective of many research projects on carbon
electrochemistry is the establishment of a relationship between
surface structure and electrochemical performance, particularly
electrode kinetics and background current. The first step
toward this goal is consideration of possible degrees of freedom
in surface structure. What are the surface structural variables
that can affect electrode performance?

1.   Distribution of Edge vs. Basal Plane on Surface

As discussed in Sec. II, bulk carbon materials vary greatly in
the density of graphitic edge plane. When a surface is prepared from a bulk material, the fraction of the surface which is
edge plane (f_e) will depend strongly on the type of carbon and
how the surface was prepared. We define f_e as the area of exposed edge plane on a particular surface divided by the total
area. This definition implies that edge plane will be intermixed
with basal plane, with $f_b = 1 - f_e$. As a trivial example, perfect basal plane HOPG has $f_e = 0.0$, while the exposed edge
plane of the same HOPG piece has $f_e = 1.0$. Given the inherent
anisotropy of graphitic carbon, one might expect different electrochemical behavior for edge vs. basal plane, with a given observation representing some weighted average of basal and edge
properties.

Even for materials with the same f_e, there may be differences
in the distribution of edge plane, with large or small regions of
edge plane being possible. For example, a graphite surface
with $f_e = 0.01$ may have many small, atomic-scale defects scattered over the entire surface (as in Ar^+ ion etched HOPG), or
the edge plane may be concentrated in one large defect. Depending on the size and spacing of the edge plane features
relative to the magnitude of $\sqrt{Dt}$, different voltammetric behavior can occur. This point will be addressed further in Sec.

III.C.2, but for the moment, it is clear that the magnitude of f_e and the size and distribution of edge plane sites are potentially important variables on carbon surfaces.

2. Roughness Factor

Except for basal plane HOPG, carbon surfaces always have some degree of roughness. We will define the roughness factor as the ratio of microscopic area to geometric area as follows:

$$\sigma = \frac{A_m}{A_g} \tag{5}$$

The microscopic area A_m is the relevant area for adsorption or kinetic measurements. For simplicity, we will assume that the surface roughness is on a distance scale which is large compared to the analyte molecules. The geometric (or macroscopic) area A_g is determined either by visual inspection or by chronoamperometry at a time scale where $\sqrt{Dt}$ is much greater than any surface roughness. In most cases, chronoamperometry will provide the more accurate A_g. The roughness factor refers to the entire microscopic area, without regard to the amount or distribution of edge plane. Thus the edge plane area for a particular electrode is given by the following equation:

$$A_{edge} = A_g \, \sigma \, f_e \tag{6}$$

σ can vary from 1.0 up, but is rarely larger than 10 for any analytically useful electrode.

3. Physisorbed Impurities

As with any other solid surface, carbon is subject to adsorption of impurities. Given the use of certain carbon materials as adsorption purifiers, carbon may be more prone to inpurity adsorption than many other electrode materials. We will designate the fractional coverage of an electrode surface by physisorbed impurities as Θ_p. As with σ, Θ_p is an average over the entire surface, without regard to whether adsorption occurs on edge or basal sites.

The identity of physisorbed impurities is generally unknown, although something can be said about the magnitude of the problem. Consider a trace organic impurity in solution with some concentration C^b and diffusion coefficient D. Further suppose that the impurity irreversibly adsorbs to the surface whenever the impurity strikes the surface. If mass transport to the surface is governed by diffusion, Eq. (7) gives the approximate time required to deposit a monolayer (5×10^{-10} moles/cm^{-2}) of adsorbate on the electrode

$$\frac{5 \times 10^{-10} \text{ mol}}{cm^2} = \frac{2 \ C^b \ (Dt)^{1/2}}{\pi^{1/2}} \tag{7}$$

where t is the time that the initially clean surface is exposed to the solution. For a typical D of 5×10^{-6} cm^{-2}/sec and $C^b =$ 1,5, and 10 μM, t is 11 hr, 2.5 hr, and 6 min, respectively. Thus, a clean electrode will be completely covered ($\Theta_p = 1$) in a solution with 5 μM impurity levels in 2–3 hr. This time will be reduced significantly by convection or higher impurity concentrations, but it is clear that a clean electrode will stay clean long enough to do several experiments. It will last much longer than the same electrode in air for similar levels of impurities, because of the much higher diffusion coefficient in the gas vs. solution phases.

4. Chemisorbed Species

Clean carbon surfaces are reactive due to the presence of unsatisfied valences and are prone to chemisorb a variety of molecules, particularly those containing oxygen. The terms chemisorbed oxygen, oxygen-containing functional groups, and surface oxides will be considered synonymous here. There is an enormous literature on the surface oxides of carbon, mainly because oxides have a large effect on surface chemistry [48].

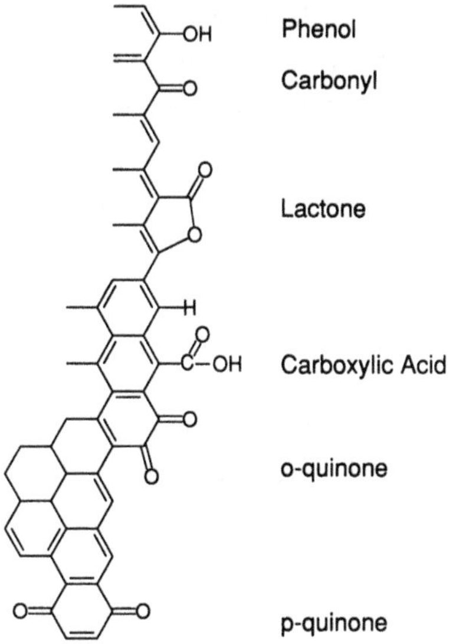

FIG. 15. Several possible oxygen containing functional groups that may be present on carbon surfaces.

A representative but not comprehensive list of surface oxide groups is shown in Fig. 15. These functional groups have been identified and in many cases quantified by infrared adsorption, wet chemical analysis, ESCA, etc. Some authors classify surface oxides as acidic, basic, and neutral, based on reactivity with known acids or bases [2,4]. Although the existence of surface oxides is well established, the type and quantity of functional groups varies greatly with carbon material and pretreatment history. As will be shown later, variations in surface oxides is a major source of variability in electrochemical performance.

The surface oxides discussed so far typically exist in sub-monolayer coverage and are located on edge plane sites. When carbon is oxidized extensively, a multilayer oxide film can result, which is a fundamentally different structure from submonolayer oxides. Graphite oxide (GO) and its electrochemically generated

analog (EGO) is a poorly defined material containing C, H, and
O in variable proportions [5]. GO has the empirical formula
C_8O_2 in fully oxidized anhydrous form, or $C_8(OH)_4$ in the hydrated form [5,49]. Full oxidation rarely occurs, however, so
empirical formulas with lower oxygen content are usually observed.
GO is often made by reaction of graphite intercalation compounds
with water, while EGO is formed when graphite is electrochemically oxidized in aqueous acids.

Although the structure of EGO is a subject of controversy,
some inferences about its structure are available. The inter-
layer spacing of graphite increases from 3.35 Å to 6.20 Å in
EGO when the graphite is oxidized to an empirical formula of
$C_{4.6}O_{0.75}$ H_2O [5]. This observation immediately distinguishes
EGO from the surface oxides of Fig. 15, since the oxygen atoms
and water are located between carbon layers, instead of on their
edges. Oxidation of graphite removes electrons from the aromatic π system, and eventually leads to aliphatic bonds in the
graphite plane. As a result, EGO has much lower conductivity
than graphite, and in some cases is an insulator. A recently
proposed structure for GO [49] is based on a hydration equilibrium:

$$C_8O_2 + 2H_2O \rightleftarrows C_8(OH)_4 \qquad (8)$$

As shown in Fig. 16, GO can be considered a layered material
with fused, saturated cyclohexyl rings with axial hydroxyl or
carbonyl groups located between the layers. As noted earlier,
the structures in Fig. 16 are idealized, fully oxidized materials;
real GO or EGO lies somewhere between these structures and
graphite.

The formation of EGO has been examined extensively in
strong aqueous acids, particularly H_2SO_4 [5,50–53]. During
anodic oxidation of the graphite, HSO_4^- intercalation compounds
are formed, in a potential range of 0.7–1.1 V vs. SCE in 96%

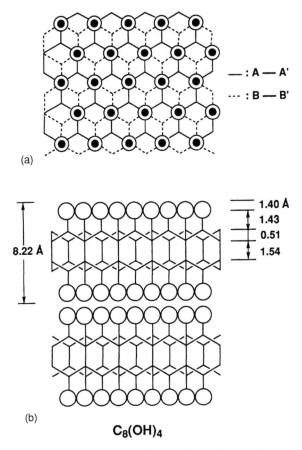

FIG. 16. Idealized structure for hydrated graphite oxide proposed by Nakajima, et al.: a) Top view of former basal plane; b) edge view. Open circles denote OH groups, closed circles in upper drawing denote OH groups below the plane of the drawing. (Adapted from Ref. 49.)

H_2SO_4. If the graphite is reduced after formation of the intercalation compound, the redox process is reversible and the original graphite is recovered [5,50]. However, if the potential
exceeds ca. 1.6 V vs. SCE, EGO forms. The intercalation of
HSO_4^- is fairly rapid, whereas EGO formation exhibits slow kinetics. Furthermore, the carbon in EGO does not reach as high
an oxidation state as in GO because the low conductivity of EGO
eventually prevents further oxidation. Raman investigations of
carbon fiber oxidation in 96% H_2SO_4 confirms that no changes
in L_a occur during the reversible formation of the HSO_4^- intercalation compound. As shown in Fig. 17, the Raman spectrum
did not change significantly through an entire oxidation/reduction
cycle, provided the applied potential remained below 1.5 V. The
intercalation process is characterized by "staging," which results
from discrete intercalated graphite structures. Figure 18 shows
several different stages of intercalation, and an idealized potential vs. time plot for galvanostatic formation of the stages. Each
jump in potential corresponds to the completion of a given stage,
while a potential plateau occurs during oxidation of the graphite
layer from one stage to the next.

After the applied potential exceeds about 1.5 volts vs. SCE
(in 96% H_2SO_4), the behavior upon reduction is quite different
from that observed for intercalation. The decrease in L_a implied by the Raman spectrum of Fig. 17 indicates that formation
of EGO irreversibly damages the graphite lattice. Maeda et al.
propose that EGO formation creates severe strain in the graphite
plane which causes it to fracture, whereas intercalation can
proceed without such strain [50]. Thus, EGO formation and
reduction is a means to create defects in basal plane regions,
while intercalation is not. This mechanism is summarized graphically in Fig. 19.

While there is uncertainty about the nature of EGO formed
in 96% H_2SO_4, the picture in less concentrated acids or neutral

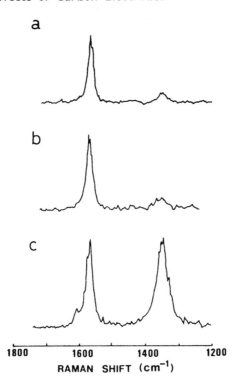

FIG. 17. Raman spectra of carbon fibers before and after intercalation cycles in 98% H_2SO_4. Spectrum a, initial; b, after
10 cycles from 0.0 to 1.4 V; c, after 10 cycles from 0 to 2.3 V.
(From Ref. 50, reprinted with permission of the Electrochemical
Society.)

media is even less clear. Intercalation should not occur in
weakly acidic or neutral solutions since the potential required
to form the intercalation compound is higher than that required
to form O_2 or CO_2 [5]. Thus, an intercalation compound should
not exist in mild acids or neutral solutions. However, lattice
damage does occur when HOPG is oxidized in 0.1 M KNO_3, based
on Raman data [31], and it is well known that an oxide film
forms on carbon electrodes under similar conditions [54–56].
The composition and nature of the oxide film formed in mildly

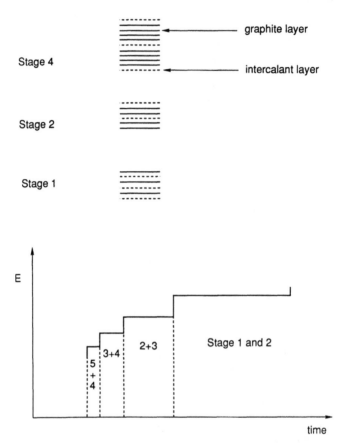

FIG. 18. Staging during constant current electrochemical intercalation of graphite. Lower trace is idealized potential vs.
time plot during galvanostatic intercalation; numbers denote
which stages are present at particular times. (Adapted from
Ref. 5.)

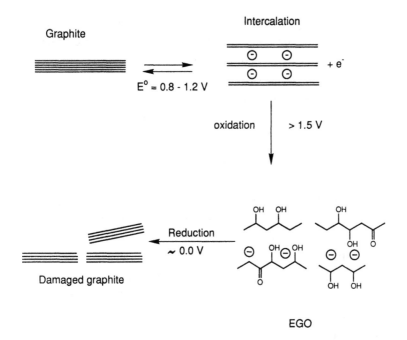

FIG. 19. Schematic of structural changes occurring during oxidation of graphite in strong H_2SO_4. Damage to graphite lattice
occurs as a consequence of EGO formation. Small circles denote
HSO_4^- ions.

acidic media has been examined, due to its importance to electrochemical activation of carbon electrodes. Engstrom and
Strasser noted that GC becomes more wettable after oxidation
at 1.8 V vs. SCE for 5 min in 0.1 M KNO_3, followed by reduction at −0.1 V [54,56] for 1 min. The contact angle before
electrochemical pretreatment (ECP) was $72 \pm 7°$, while that
after was $54 \pm 6°$. The decrease in contact angle implies a
more hydrophilic surface, as might be expected for a hydrated
EGO film. Auger analysis of the film indicated an increase of

the O/C ratio from 0.084 before ECP to 0.24 after, confirming the formation of an oxygen rich film [54]. Kepley and Bard [55] conducted elemental analysis of a sample of oxide film generated from GC after several weeks of potential cycling from 0.24 to 2.04 V vs. SCE at 0.2 V/sec. After vacuum drying, the elemental analysis yielded an empirical formula of $C_{3.6}H_{0.75}$ $OS_{.009}$. The negligible sulfur content implies that very little HSO_4^- remained in the film, and the C, H, and O contents are within the range observed for EGO formed in a stronger acids. Kepley and Bard also demonstrated that the observed d_{002} spacing increased from 3.43 to 4.14 Å during oxidation. Furthermore, their measurements of optical properties indicated that the bulk of the oxide film is nonconductive. All of these observations on oxide films generated in mildly acidic or neutral solution imply that the oxide film is similar to EGO formed in strong acid, except for the apparent lack of intercalation of electrolyte anions.

To summarize the qualitative picture of EGO, it is a hydrated, porous, anionic film with a thickness up to at least a 1 μm. The graphite planes of the original carbon are delaminated and their L_a is reduced. The degree of oxidation of carbon in EGO can vary over a significant range, presumably depending on electrolysis conditions. In strong acid, EGO is reducible back to graphite, albeit with smaller L_a.

When considering either submonolayer coverage of oxygen functional groups or thick oxide films, both the identity and the relative abundance of different groups may vary depending upon surface preparation and sample history. Unfortunately, one rarely knows these important characteristics with sufficient accuracy to predict the effect of surface oxides on electrochemical behavior. Nevertheless, valuable correlations of oxide film characteristics with electrochemical observable are available (see Sec. IV).

5. Separation of Variables on Carbon Electrode Surfaces

Even in the absence of surface oxides, there is enough variability of carbon surfaces to frustrate attempts to characterize them. As an example, consider an experiment designed to determine the microscopic area of a carbon electrode, A_m. A straightforward procedure involves the voltammetry of a redox molecule which strongly adsorbs on the surface. For example, phenanthrenequinone, PQ, forms a monolayer film on the electrode surface, and the moles of adsorbed PQ may be determined voltammetrically or from chronocoulometry [57] via Eq. (9):

$$Q_{ad} = nFA_m \; \Gamma^S_{PQ} \tag{9}$$

where

Q_{ad} = charge due to adsorbed PQ
Γ^S_{PQ} = saturation coverage of PQ, moles/cm^2
n = 2

Γ^S_{PQ} can be predicted from a knowledge of the molecular area, as has been done extensively for platinum surfaces [58].

Assuming that PQ covers the entire microscopic area, the roughness factor is eaily determined from A_m and A_g:

$$\sigma = \frac{A_m}{A_g} = \frac{Q_{ad}}{nF \; \Gamma^S_{PQ} A_g} \tag{10}$$

A_g is usually determined from chronoamperometry, as noted earlier. The assumption of even coverage over the carbon surface may be correct, but it is also possible that PQ only adsorbs on clean edge plane sites which are exposed at the surface. Thus, any basal plane or physisorbed impurities may inhibit PQ adsorption, leading to a smaller Q_{ad}. In the case of a partially blocked ($\theta_p > 0$) electrode with f_e coverage of edge plane, the area available for adsorption will equal $A_m f_e (1 - \theta_p)$, and

$$Q_{ad} = nF \; \Gamma_{PQ}^{s} \; A_M \; f_e \; (1 - \Theta_p) \qquad\qquad (11)$$

$$\Gamma_{obs} = \frac{Q_{ad}}{nFA_g} = \Gamma_{PQ}^{s} \; \frac{A_m}{A_g} \; f_e \; (1 - \Theta_p) \qquad\qquad (12)$$

Note that Γ_{obs} and Q_{ad} no longer depend only on σ, but also on f_e and $(1 - \Theta_p)$, and one actually determines the product $f_e \; \sigma \; (1 - \Theta_p)$ from voltammetry. If one were trying to use PQ adsorption to determine roughness changes caused by some pre-treatment procedure, it would be difficult to determine whether σ, f_e, or Θ_p were changing during the procedure. Since adsorption, electron transfer rate, and capacitance are all expected to depend on A_m, the problem of separation of surface variables often arises when studying carbon. There is usually more than one variable which affects the observed capacitance, adsorption, or heterogeneous rate constant and one or more variables are often uncontrolled and/or unknown. It is not surprising that the electrochemical performance of carbon electrodes varies so greatly. A schematic representation of surface variables of relevance to electrochemistry is shown in Fig. 20.

B.   Structural Characterization of Carbon Surfaces

Due to their importance to adhesion, adsorption, and reactivity, the surfaces of carbon materials have been the subject of major research efforts. A wide variety of techniques have been used to characterize carbon surfaces, for samples ranging from powders such as carbon black and powdered graphite, to bulk solids such as glassy carbon disks and coal. Since the subject here is electroanalytical applications of carbon, several methods used to examine carbon electrode surfaces will be discussed. A comprehensive discussion of techniques will not be attempted, and the choice of methods covered is based on their relevance to the characterization of carbon electrodes. Probes of surface

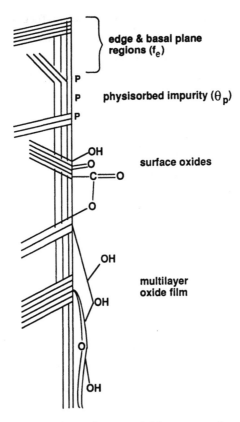

edge & basal plane regions (f_e)

physisorbed impurity (θ_p)

surface oxides

multilayer oxide film

FIG. 20. Schematic of surface variables on carbon surfaces. See text for details.

structure and composition will be considered first, with electro-chemical characterization to follow.

1. Ultrahigh Vacuum Surface Techniques

Of the many high vacuum surface science techniques, Auger electron spectroscopy (AES), and electron spectroscopy for chemical analysis (ESCA, also known as x-ray photoelectron spectroscopy, XPS) are the most common for examining carbon electrode surfaces [59–69]. Both have a sampling depth of a few atomic layers and are therefore more surface selective than most optical techniques. Both AES or low resolution ESCA have been used to determine the elemental composition of carbon surfaces, with detection limits of about 1 atom%. Figure 21 shows an example of low resolution ESCA for determination of surface oxygen content. In order to determine the surface O/C atom ratio, the raw spectrum must be corrected for the sensitivity of the instrument for different elements. From Fig. 21, it is clear that heat treatment greatly reduces the surface O/C ratio

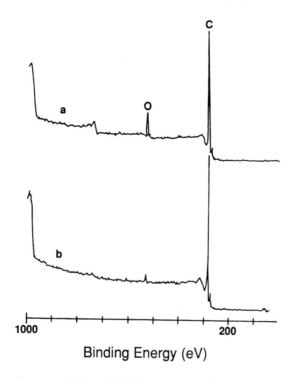

Binding Energy (eV)

FIG. 21. Low resolution ESCA spectra of GC-20 before and after heat treatment. a) Polished surface; b) after vacuum heat treatment at 700°C and <10^{-6} torr. (Adapted from Ref. 59.)

[59]. Survey ESCA spectra are especially useful for monitoring changes in surface composition during electrode pretreatment or derivatization.

When the C_{1s} peak in the ESCA survey spectrum is examined at high resolution, fine structure is observable which is attributable to covalent bonding of the carbon to other elements, notably oxygen. In pure carbon, the graphitic carbon ESCA peak is at 284.3 eV [60], and there is a long tail toward higher binding energy which has been attributed to interaction with the conduction band of the graphite [61]. The magnitude of the tail varies with the state of surface derivatization, since it will affect the conduction band electrons. Superimposed on this tail are chemically shifted C_{1s} peaks which appear at binding energies above 285 eV because of covalent bonding to oxygen. Resolution of these oxide peaks is difficult, since the peak widths are comparable to the chemical shifts and they are superimposed on a gradually declining tail of the graphitic carbon peak. In addition, there is some variation in the reported peak positions due to instrumental variables. Figure 22 shows the effect of electrochemical oxidation on the C_{1s} peak of carbon fibers. The solid lines show the contributions of three chemically shifted C_{1s} peaks to the high binding energy tail. Table 5 lists chemical shift values and the assignments to different surface oxides [60–67]. Although there is some uncertainty about accurate chemical shift values and assignments, there appears to be general agreement that the shifted peaks are attributable to alcoholic, C=O, carboxylate, and carbonate species. By various curve fitting procedures, the contribution of each of these surface functional groups may be assessed. For example, Fig. 22 was used to conclude that oxidation of a carbon fibers in 0.22 M HNO_3 resulted in principally carbonyl/quinone and ether groups on the surface [61]. Several additional conclusions from ESCA experiments will be presented when particular carbon materials are discussed later.

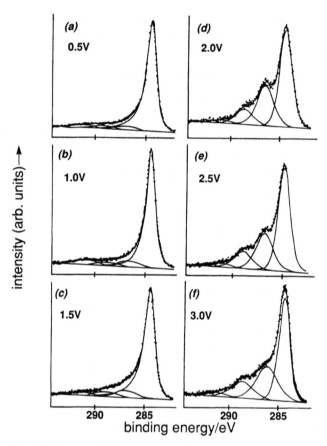

FIG. 22.   High resolution ESCA spectra of carbon fibers subjected to electrochemical oxidation.   Fibers were held at the indicated potentials for 20 min in 0.22 M HNO$_3$.   Smooth curves
resulted from curve fitting to various distributions of surface
oxides.   (Adapted from Ref. 61.)

TABLE 5

Chemical Shifts for C_{1s} XPS Peak

Carbon type	Chemical shift, eV, relative to 284.3 eV
Graphite	0
C–H	0.8
Alcohol, phenol	1.6–2.1
Carbonyl, quinone	2.1 – 3.5
Carboxylate, ester	4.0 – 4.8
CO_3^{-2}	>6.0

Source: From Refs. 60–67.

In order to avoid the uncertainties associated with curve fitting of high resolution ESCA spectra, the surface can be derivatized with a functional group specific tag before obtaining low resolution spectra. For example, surface carbonyl groups react with pentafluorophenyl hydrazine (PFPH) introducing F and N onto the surface.

$$-C=O + C_6F_5N-NH_2 \rightarrow -C=N-NC_6F_5 + H_2O \qquad (13)$$

A survey ESCA spectrum will then show the F or N markers, without requiring a high resolution spectrum. Provided the process can be calibrated, the F/C or N/C ratios will provide an estimate of surface carbonyls [68]. The approach has also been applied to determining surface phenols [69]. The results vary with sample preparation, but surface phenols on GC were reported as 1–9%, with about 2% of the surface being carbonyl groups.

2. Surface Raman and Infrared Spectroscopy

As noted earlier, Raman spectroscopy has a sampling depth of a few hundred Å. This depth is sufficient to probe carbon microstructure near the surface, but the data will be less

surface selective than that from ESCA, which samples about 10
Å. Surface monolayers have been observed with Raman spectroscopy on metal surfaces in high vacuum [70], but not on carbon
as yet. There is one example in the literature on surface enhanced Raman of carbon, obtained by vapor depositing small
Ag silver islands onto graphite or glassy carbon [71]. The
results show that Raman may be made more surface selective,
since the SERS sampling depth was about 20 Å. However, neither Raman nor SERS has yet been used to obtain spectra of
monolayers on carbon surfaces. An additional impediment for
using Raman to probe surface monolayers is the weak scattering
cross section of most of the surface oxygen functional groups.

Infrared absorption has higher sensitivity than Raman for
surface oxides on carbon, partly because the dipole moment
change of many surface groups is stronger than the polarizability change. IR has been used fairly extensively to characterize
high surface area carbon materials [72]. An increased area
provides more sample for a given surface concentration, thus
achieving a useful S/N for IR absorption spectra. Such experiments have identified numerous surface groups on powders,
and have been used to monitor surface derivatization [73] and
to confirm conclusions from ESCA on bundles of carbon fibers
[61]. Since high surface area methods are rarely useful for
electrodes used in electroanalysis, they will not be discussed
further.

When IR is applied to relatively smooth electrode surfaces,
the sensitivity problem is severe. Since a typical monolayer is
ca. 5×10^{-10} moles/cm^{-2}, very little sample is in the beam.
This issue has been examined in detail for metal electrodes [74,
75], and several points from that work are relevant. First, the
high reflectivity of metals (e.g., Pt and Au) in the IR results
in a surface selection rule. Light polarized parallel to the electrode surface has a negligible electric field at the surface, due

to formation of an image dipole which destructively interferes with the incident light. Incident light polarized perpendicular to the surface is enhanced by the image dipole, however, thus favoring absorption by molecules oriented with their dipole moments perpendicular to the surface. Thus only vibrations perpendicular to the surface absorb IR light polarized perpendicular to a metal electrode. This selection rule is weaker for carbon electrodes because of lower reflectivity and weaker image dipole [76]. As a consequence, modulation techniques used in surface IR spectroscopy of metals are less effective on carbon, and to date IR spectra of surface oxides on carbon have not been obtained. However, the optical properties of carbon can be used to advantage when observing thin polymeric films [76].

A final lesson from IR examination of metal electrode surfaces is the high value of surface vibrational information. Since ESCA is a probe of elemental composition and oxidation state, it provides relatively little information about molecular structure. We will see later how useful Raman vibrational data can be for carbon electrodes, even though the sampling depth is much greater than a monolayer. Thus, surface IR on metals and Raman on carbon surfaces provide molecular information about electrode processes which is not available from high vacuum or other techniques. In addition, Raman or IR can often be used in situ, avoiding the uncertainties associated with sample transfer to high vacuum. Although Raman and IR techniques have not yet developed to a point where they can be applied to monolayers on carbon electrodes, there is no fundamental reason why they should fail. As will be discussed later, vibrational information provides crucial clues about the electrochemical behavior of carbon materials without requiring monolayer sensitivity.

3. Contact Angle and Wettability

It has long been recognized that surface derivatization can affect the wettability of carbon electrodes. In simple terms, the contact angle ϕ of a droplet on a flat surface is defined as shown in Fig. 23. A hydrophobic surface leads to a nearly hemispherical droplet, and the contact angle approaches 90° (cos ϕ → 0). A hydrophilic surface attracts the liquid, and is "wettable", with ϕ approaching 0° (cos ϕ → 1). Thus an increase in cos ϕ indicates a more hydrophilic surface. Carbon surfaces which have many oxygen-containing functional groups generally have smaller contact angles, while those without oxides have higher ϕ. Although contact angle provides a measure of wettability, it is indirectly coupled to surface structure, and changes in contact angle do not necessarily indicate surface oxidation. For example, HOPG basal plane exhibits a large decrease in ϕ for water following ion bombardment in vacuum, with no observed surface oxides [77]. The contact angle decreased from 35° to 0° when HOPG basal plane was bombarded with Ar$^+$ in UHV, and surface oxygen was undetectable (<1%) by AES before or after Ar$^+$ treatment, or after the contact angle measurement.

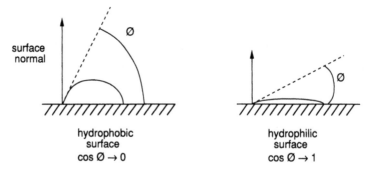

FIG. 23. Definition of the contact angle, ϕ, between a droplet of liquid and a solid surface.

4. Scanning Tunneling Microscopy (STM)

Many of the structural features of carbon materials, e.g., L_a and d_{002}, have magnitudes of less than 100 Å for electrochemically useful materials. These features are below the resolution limit of scanning electron microscopy, and most electrodes are much too thick to study by high resolution transmission electron microscopy. The atomic scale resolution of STM, plus its compatibility with electrolyte solutions, provides significant motivation to apply STM to carbon electrodes. Even when the electrode is immersed in solution, STM can provide topographic information with a resolution of a few Å. Figure 24 shows an STM of fresh HOPG basal plane in air and in solution [78]. Upon oxidation of the HOPG surface by potential cycling between 0.0 and 1.8 V vs. a silver quasi reference electrode, damage to the surface is apparent (Fig. 25). Note that the major feature evident in Fig. 25C is about 100 Å wide, and would barely be observable by SEM. Furthermore, STM can provide an estimate of tunneling barrier height and therefore some insight into the surface electronic structure. It is likely that STM and its electrochemical analog scanning electrochemical microscopy [79–81] will be frequently used to study carbon surfaces in the future.

5. Thermal Desorption Mass Spectrometry

Although TDMS has been used to characterize carbon surfaces for some time, its application to electrode materials is quite recent. Of particular interest is the relationship between TDMS behavior and surface oxygen functional groups which may be important to electrochemical behavior. Fagan and Kuwana [82] used TDMS to examine carbon fibers before and after heat treatment and gas phase oxidation. They monitored primarily CO and CO_2 as functions of fiber temperature. CO would be expected to desorb from surface phenol, carbonyls, or quinones, while CO_2 should result from lactones or carboxylates. After

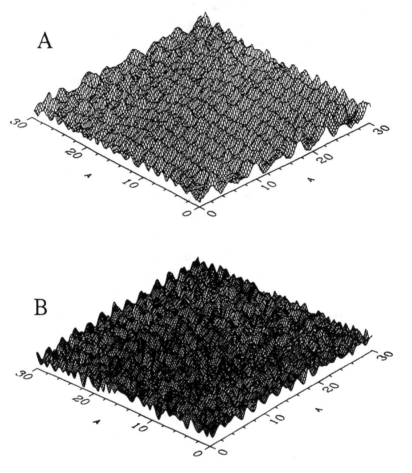

FIG. 24. STM images of HOPG in air (A) and 0.1 M H_2SO_4 (B). (From Ref. 78, reprinted with permission.)

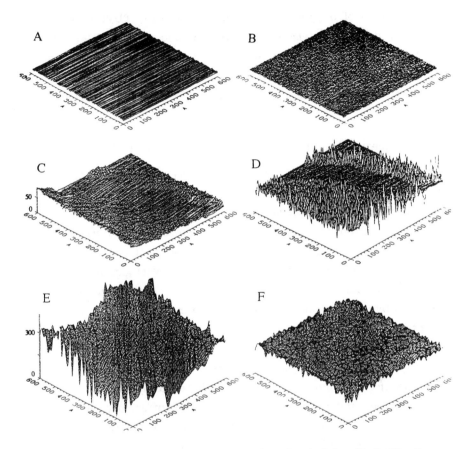

FIG. 25. STM images (A,C,E) and barrier height (B,D,F) of HOPG in 0.1 M H_2SO_4. A,B are before oxidation; C,D are after 20 oxidation/reduction cycles; and D,E are after further oxidation. (From Ref. 78, reprinted with permission.)

chemisorption of oxygen at elevated temperature, TDMS exhibited four desorption temperatures for CO and two for CO_2. The
desorptions were tentatively assigned to various surface groups.
As with STM, TDMS has been demonstrated to be feasible and
shows significant promise for the future.

C. Electrochemical Characterization of Carbon Electrodes

The stated goal of this chapter is the correlation of surface
structure with electrochemical behavior. In this section, several electrochemical observations which are affected by surface
structure will be discussed. No attempt to be exhaustive will
be made, rather emphasis is placed on those electrochemical effects which are commonly monitored during investigations of
carbon electrode structure and preparation. These include
voltammetric background, electron transfer kinetics, and adsorption.

1. Voltammetric Background Current

Voltammetric techniques have very high sensitivity in terms of
response per unit of concentration. Stated differently, easily
measured currents correspond to very low concentrations, e.g.,
one picoamp of oxidation current is produced by 10^{-17} moles of
reduced species oxidizing per second. Unfortunately, this sensitive response is superimposed on a relatively large background
current for most voltammetric methods. For example, a typical
metal electrode (0.1 cm^2) has a double layer capacitance C_{dl} of
20 $\mu F/cm^2$ and will exhibit a background charging current of
0.2 μA at a scan rate of 0.1 V/sec. This background current
ultimately establishes the limit of detection for linear sweep
voltammetry at values of ca. 10^{-5} M. Thus capacitive background
current can be a significant interference when using voltammetry
for trace analysis.

The capacitive component of the voltammetric background has long been used on Hg electrodes to study electrode/solution interfaces, particularly the effects of adsorption, electrolyte concentration, etc., on double layer capacitance. The overall objective of these experiments is an understanding of double layer structure and its effects on electrode kinetics. Similar techniques have been applied to carbon electrodes, but unfortunately the carbon/solution interface is much less well behaved than the mercury/water interface. The surface roughness, surface functional groups, and possible surface faradaic reactions on carbon electrodes greatly complicate the observed "background current" and usually frustrate attempts to determine surface structure from voltammetric background.

Based on a variety of examinations of background current, there are at least two and possibly three mechanisms responsible for voltammetric background current at carbon electrodes. First, carbon electrodes should have a double layer capacitance similar to that of other solid electrodes. This capacitance will be denoted C_{dl}^o, with units of microfarads per cm^2 of geometric area. The superscript "o" will be used here to denote a capacitance normalized to the geometric area (A_g) defined in the same fashion as in Eq. (5). While C_{dl}^o can vary significantly for various electrode and solution conditions, it is usually 10–30 $\mu F/cm^2$ for solid electrodes in aqueous electrolytes. A second contribution to background is due to surface faradaic reactions, particularly quinones. Since a variety of quinones with different E^o may be present, the faradaic background current may be weakly potential dependent, and may appear similar to charging current. For this reason, surface faradaic reactions are sometimes said to cause "pseudocapacitance." It is often possible to observe a pH-dependent voltammetric couple with a potential of 0.2–0.3 V vs. SCE at pH 2 [83]. These reactions will be discussed later, but it is clear that they produce voltammetric currents which have a mechanism distinct from double layer capacitance.

While capacitive and faradaic components of the voltammetric background are well accepted, a more controversial third mechanism for background current has been proposed. The experimental observation is a "slow" current response to a potential change, lasting many msec or sec [84–86]. Since double layer charging should require less than 1 msec in most aqueous experiments, the "slow capacitance" is anomalous. There are several possibilities for the origin of "slow capacitance." It could be double layer charging, but with an RC controlled by the high resistance inside pores or microcracks in the carbon surface. Ordinary double layer charging would be slower for a carbon surface inside a crack, due to a large $R_s C_{dl}$ time constant. For a variety of cracks, a variety of RC values would occur leading to slow charging. A second possibility is a surface faradaic reaction which is very slow, or one occurring inside a pore. Nagaoka et al. [87–89] propose reaction 14 to explain anomalous background currents.

$$>C=O + H^+ + e^- \rightleftarrows >COH \tag{14}$$

They propose that H^+ transport into pores or cracks may be slow, or the reduction itself is slow, thus yielding the observed "slow" current. Finally, superficial intercalation has been proposed [84–86], in which redox reactions of the graphitic system occur, with accompanying ion transport. This process should be slower than double layer charging, and might occur over a broad potential range. Regardless of the origin of "slow capacitance," it is clear that background currents attributable to capacitance or surface redox process occur on carbon surfaces, and they can vary widely in potential and time dependence.

Before temporarily leaving the subject of background currents, we will briefly discuss several common measurements of background current, particularly the ways that surface events are manifested in background response. Throughout the discussion C_{dl}^o will be used to denote the true double layer capacitance in $\mu F/cm^2$, and C_{obs}^o will denote the observed "capacitance," regardless of its origin. Obviously C_{obs}^o will depend on the time scale and potential of the method employed. It is practical to classify various measurements of background by their time scale and magnitude of the potential perturbation.

a. Fast, Small Amplitude AC Techniques

At the limit of high frequency, the current response to an imposed potential waveform should reflect only charging current, i.e.:

$$i_{ch} = C_{dl}^o \, A_g (dE/dt) \tag{15}$$

where dE/dt is the instantaneous rate of potential change. If dE/dt is high, then the faradaic or "slow" capacitive currents should be minimized, and the observed capacitance should approach C_{dl}^o. A convenient variant of this AC approach was presented by Gileadi [90,91], in which the potential waveform was a small amplitude triangular wave. For this case, the current predicted from Eq. (8) approaches a square wave of amplitude $AC_{obs}^o \, \nu$, where ν is the slope of the triangular wave in V/sec. Since the amplitude of the perturbation is small, faradaic contributions to C_{obs}^o are diminished. The current waveforms for such an approach are shown in Fig. 26 for an HOPG basal plane electrode [92]. Note that there is a slight slope to the top of the square wave on a laser activated surface. However, the great majority of the current transient is very fast, supporting the conclusion that the method is measuring primarily C_{dl}^o. This method has some similarities to the methods used by Soffer et al. [84–86] to discriminate "fast" and "slow" capacitance.

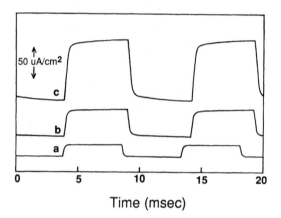

FIG. 26. Current vs. time waveforms resulting from 100 Hz, 20 mV p-p triangular potential waveform. (From Ref. 92, reprinted with permission.)

b. Small Amplitude Chronocoulometry

Fagan, Hu, and Kuwana (59) used chronocoulometry to determine the charge required to change the potential of a GC surface, with:

$$C^o_{obs} = \frac{Q}{\Delta E \, A_g} \tag{16}$$

where Q is the observed charge, A_g the area, and ΔE the size of the potential change. As ΔE decreases, Q should contain a smaller contribution from faradaic processes. Since the approach is integrative and slower than most AC techniques, it would be expected to have a larger faradaic or "slow capacitive" contribution than the triangular wave techniques just described. Results of the approach are shown in Fig. 27. The expected linear dependence of Q on ΔE is observed at small ΔE, and the curvature at higher ΔE was attributed to faradaic processes. It should be emphasized that Q (and C^o_{obs}) will depend on the integration time used to determine Q.

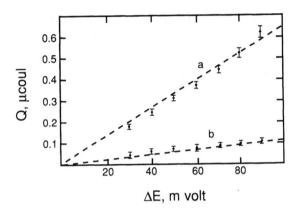

FIG. 27. Charge resulting from 1-sec long potential steps if height ΔE on GC-20 electrodes in phosphate buffered (pH 2) KCl. Curve a is a polished electrode, b is after VHT. (Adapted from Ref. 61.)

c. Linear Sweep Cyclic and Differential Pulse Voltammetry

Linear sweeps cyclic voltammetry (CV) is the most common technique used for examining carbon background, almost by default, since it is the most common method employing carbon electrodes. As with all measures of background on carbon, the CV response is a strong function of electrode history. The CV results for various pretreatments will be discussed later, but for now, several features of the CV background deserve note. First, the background current is usually larger than that expected from simple double layer charging. For example, the background current observed by Fagan, Hu, and Kuwana for a polished GC surface [93] corresponds to a C_{obs}^{o} of ca. 120 $\mu F/cm^2$. Much of this apparent capacitance was attributed to surface faradaic reactions. Since the majority of CV experiments are conducted at moderate scan rates (10–200 mV/sec), there is sufficient time for faradaic reactions or "slow capacitance" to contribute to the current.

Second, CV background often shows a pH-dependent quasi-reversible couple near 0.2 V. Its current is proportional to scan rate, indicating it is a surface-bound redox couple. Nagaoka et al. [89] note that the peak potential of the couple varies linearly with pH at a rate of −60 mV/pH unit, indicating that the redox process involves an equal number of protons and electrons. The surface-bound peaks are more readily apparent in a differential pulse voltammogram of a polished GC surface [59,83] (Fig. 28). Both low potential (ca. 175 mV vs. Ag/AgCl) and high potential (700 mV vs. Ag/AgCl) peaks were observed

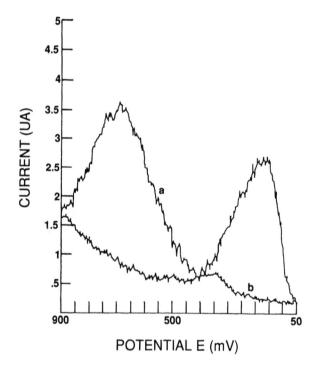

FIG. 28. Differential pulse voltammograms of GC-20 electrodes in pH 2 phosphate buffer. Curve a is polished, b is a VHT surface. (Adapted from Ref. 61.)

at pH 2, and these peaks have been assigned to surface orthoquinone groups by several authors [59,83,94], although alternative explanations have been proposed. Nagaoka argues that the
proposed "quinone" peak may in fact be caused by the reaction
in Eq. (14), rather than surface quinones. Nonetheless, authors
agree that the voltammetric background peaks are related to surface oxygen groups.

In summary, the background currents on carbon can be
distinguished as true capacitive charging current and as surface
faradaic current. Ideally, capacitive current should behave like
a capacitor of ca 20 $\mu F/cm^2$ and a time constant equal to $R_s C_{dl}$.
Ideal surface faradaic currents should be distinguishable as
voltammetric waves with some definable E^o. Unfortunately carbon surfaces do not behave ideally and the background current
is anything but well defined. For small potential excursions
and short measurement times, an ideal charging current is approached. At longer times, a faradaic contribution is usually
unavoidable. To make matters worse, a slow component of the
background current is often present and may be capacitive or
faradaic in origin. Despite these complexities, background current can provide important information about surface structure,
as will be apparent in Sec. IV.

2. Electron Transfer Kinetics

The driving force behind much of the research cited herein is
the examination of electron transfer (ET) rates on carbon electrodes. The practical benefits of improvements in ET rates on
carbon include enhanced resolution and sensitivity of analytical
sensors, and in the long term, applications of carbon electrodes
in energy conversion devices such as fuel cells. On a more
fundamental level, many investigators seek to understand ET
on a molecular level. If relationships between surface structure
and ET rates can be found, insight into reaction mechanisms
and kinetics will be available. Such structure–activity

relationships are difficult to establish on carbon electrodes because of the uncertainty about surface structure and stability.
On the other hand, the diversity of surface chemistry on carbon
permits manipulations of the surface that are not readily available
on metals. For example, the surface may be modified by covalent
attachment of reactive molecules [95]. Before discussing the
many factors affecting ET rates on carbon, we will discuss several redox systems commonly used to test ET on carbon. These
redox systems are conveniently classified into three groups:
quasi-reversible inorganic systems, quasi-reversible organic
systems, and chemically irreversible systems.

Throughout this chapter the nomenclature of Bard and
Faulkner [96] will be used. The standard heterogeneous ET
rate constant, k^o, is valid at the $E^{o'}$ of the system of interest,
such that Eq. (17) applies. All rate constants discussed are
uncorrected for double layer effects.

$$i_{cathodic} = nF \ A \ C^o_{ox} \ k^o \ exp\left[\frac{-\alpha nF}{RT} (E_{app} - E^{o'}) \right] \quad (17)$$

where:

C^o_{ox} = surface concentration of Ox

α = transfer coefficient

E_{app} = applied potential

a. Quasi-reversible Inorganic Redox Systems

During the many investigations of solid electrode behavior, most
investigators sought an uncomplicated outer sphere redox system
to test the effects of surface preparation, etc. If the redox
system were well behaved, then observed variations in k^o could
be attributed to changes in the electrode surface. Common
candidates in aqueous experiments include $Fe(CN)_6^{-3/-4}$, Ru
$(NH_3)_6^{+2/+3}$, $Ru(bpy)_3^{+2/+3}$, and $IrCl_6^{-2/-3}$. $Fe(CN)_6^{-3/-4}$ is by
far the most commonly studied for carbon electrodes in aqueous

media, with the other three being used much less frequently. Since $Fe(CN)_6^{-3/-4}$ has become a benchmark for carbon ET kinetics, it will be discussed in some detail, followed by brief comments on the other systems.

ET kinetics of $Fe(CN)_6^{-3/-4}$: At first glance, $Fe(CN)_6^{-3/-4}$ in 1 M KCl is an uncomplicated quasi-reversible redox system which has been used for decades to calibrate instrumentation, determine electrode area, etc. [97–105]. The standard potential is 0.120 V vs. SCE, $D_{ox} = 7.63 \times 10^{-6}$ cm^2/sec, $D_{red} = 6.32 \times 10^{-6}$ cm^2/sec [98], and $\alpha = 0.5$. Goldstein and Van De Mark compiled k° values for $Fe(CN)_6^{-3/-4}$ on platinum [99] and studied a variety of electrode cleaning procedures, with some of the results shown in Table 6. Although the cleaning procedure strongly affects the observed k°, the highest values from different laboratories are in the range 0.10–0.24 cm/sec. The need for caution on surface preparation is obvious, since a range of more than a factor of 20 in k° was observed for different preparation procedures. It is interesting to note that the maximum k° values observed on Au and Pt agree within a factor of 2, as one would expect for an outer sphere process. Furthermore, Galus et al. observed that intentional chemisorption of iodine on the Pt electrode caused a very slight decrease in the observed k° [103].

Upon closer examination, $Fe(CN)_6^{-3/-4}$ has some complications which cast doubt on its classification as a simple outer sphere redox system. The formal potential, $E^{o'}$, is quite dependent on the identity and concentration of the electrolyte [101]. For example, $E^{o'}$ varies from 0.19 to 0.26 V vs. SCE when the LiNO$_3$ concentration varies from 0.2 to 2.0 M. This may be an activity effect due to the high charges on the $Fe(CN)_6^{-3/-4}$ species, or it may indicate association of the $Fe(CN)_6^{-3/-4}$ with one or more cations. As shown in Table 6, the observed k° is strongly dependent on the electrolyte cation, with an almost linear dependence on [K$^+$] [101]. The k° is

TABLE 6

$k°$ for $Fe(CN)_6^{-3}/-4$ on Pt Electrodes

Pretreatment	Electrolyte	$k°$ (cm/s)	Ref.
Aqua regia, anodization, flame	1 M KCl	0.23 ± 0.07	99
Nitric acid, flame	1 M KCl	0.23 ± 0.23	99
Flame only	1 M KCl	0.22 ± 0.01	99
Nitric acid, anodization	1 M KCl	0.11 ± 0.04	99
Nitric acid	1 M KCl	0.02 ± 0.007	99
Aqua regia	1 M KCl	<0.01	99
$HClO_4$, redn in KCl	1 M KCl	0.24	100
$HClO_4$ only	1 M KCl	0.028	100
Potential cycling	1 M KCl	0.10	101
Potential cycling	1 M $LiNO_3$	0.01	101
Aqua regia, cathodization, flame	1 M KCl	0.21	99
Aqua regia, cathodization, flame	1 M NaCl	0.17	99
Aqua regia, cathodization, flame	1 M LiCl	0.07	99
Aqua regia, cathodization, flame	1 M $NaClO_4$	0.15	99
Potential cycling	1 M KCl, 0.01 M NaCN	>0.1	103
Potential cycling	1 M LiCl	0.08	103
Potential cycling	1 M LiCl	0.002	103
Gold electrode	1 M KCl	0.10	102

smaller for Li^+, and varies in the order $Li^+ < Na^+ < K^+ < Cs^+$
[101]. There is a weaker but still significant effect on the
identity of the anion, with k^o following the trend $I^- > Br^- >$
$Cl^- > F^-$ [99]. In 1 M KBr, a k^o of 0.31 ± 0.07 cm/sec was
observed [99]. Finally, the $Fe(CN)_6^{-3/-4}$ redox system forms
a film on Pt electrodes [103], which appears to result from the
loss of CN^- and the formation of compounds similar to Prussian
blue. This film can seriously reduce k^o, but its formation can
be inhibited by the addition of excess CN^- or chemisorption of
iodine on the Pt [103]. In the presence of 0.01 M NaCN, Kawaik,
Kulesza, and Galus measured reproducibly high k^o values, while
in the absence of CN^-, the k^o values were much lower and decreased with time. Stability of $Fe(CN)_6^{-3/-4}$ was reduced when
K^+ was absent [103].

The mechanism(s) of the cation and anion effects on k^o is
not completely clear, but several hypotheses have been made for
Pt electrodes. K^+ may form a bridge between $Fe(CN)_6^{-3/-4}$
and the Pt surface, thus providing a fast inner-sphere ET
route [104]. If so, the bridge is not affected by the adsorption
of iodine, and any mechanism involving $Fe(CN)_6^{-3/-4}$ adsorption
appears very unlikely. Peter et al. [101] concluded that the
transition state must contain one or more cations, but the cation may be merely associated with the redox center and not
necessarily act as a bridge. The anion effect appears to result
from chemisorption of the anion to the Pt surface, with the irreversible chemisorption of the anion to the Pt surface, with
the irreversible chemisorption of iodide being the most severe
case. The anion may keep the Pt surface clean by resisting
further adsorption, since the anion effect on k^o follows the
same trend as the strength of adsorption [99].

These complications have not been examined for carbon
electrodes because of the difficulty in preparing a reproducible
surface. In terms of providing a benchmark for carbon electrode

performance, most workers have used $Fe(CN)_6^{-3/-4}$ in 1 M KCl, without excess CN^-. The observed k^o values are reproducible provided the electrode surface is prepared reproducibly. It will be noted later that carbon electrodes can exhibit k^o values for $Fe(CN)_6^{-3/-4}$ of greater than 0.2 cm/sec in 1 M KCl, and that k^o is stable for hours with proper care.

Attempts have been made to predict the outer-sphere hetero-genous rate constant for $Fe(CN)_6^{-3/-4}$ from Marcus theory via Eq. (18) [105].

$$k^o = 3 \times 10^{-2} \sqrt{k_{homo}} \tag{18}$$

where k_{homo} is the homogeneous self exchange rate constant. Equation (18) predicts a k^o for $Fe(CN)_6^{-3/-4}$ in 1 M KCl of above 1 cm/sec [105]. Despite concerns about film formation or cation effects, $Fe(CN)_6^{-3/-4}$ in 1 M KCl is a useful benchmark for carbon surface preparation, provided one is aware of pos-sible complications. Based on the observation of high, stable k^o values for $Fe(CN)_6^{-3/-4}$ on carbon electrodes, degradation or film formation is not apparent.

$\underline{Ru(NH_3)_6^{+2/+3}, \; IrCl_6^{-2/-3}, \; and \; Ru(bpy)_3^{+4/+3}}$: These redox couples are believed to undergo fast, outer-sphere, one-electron ET reactions at solid electrodes. Like $Fe(CN)_6^{-3/-4}$, the cyclic voltammetry of these three systems shows well-behaved quasi-reversible waves with different E^o values. The E^o values vs. SCE have been reported as follows: $Ru(NH_3)_6^{+2/+3}$, -0.260 V vs. SCE [106]; $IrCl_6^{-3/-2}$, 0.741 V vs. SCE [106], and $Ru(bpy)_3^{+2/+3}$, $+1.03$ V vs. SCE [107]. The examination of ET kinetics for these systems is much less extensive than those of $Fe(CN)_6^{-3/-4}$.

b. Quasi-reversible Organic Redox Systems

Since carbon electrodes have been used extensively in studying the electrochemistry of organic compounds, a wide range of

compounds has been used to assess electrode behavior. Of the
quasi-reversible systems studied, the hydroquinone/paraquinone
and catechol/orthoquinone systems are most commonly used to
assess carbon electrode performance. Of particular note is
dopamine (DA) (Fig. 29), which has been studied extensively
because of its importance to neurochemistry. Activation of carbon electrodes separates the DA oxidation wave from that for
ascorbic acid, the most serious interferant in tissue. This
phenomenon will be discussed later, but for now we will consider the oxidation mechanism of dopamine and related quinones.

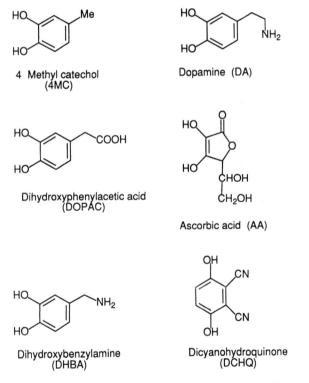

4 Methyl catechol
(4MC)

Dopamine (DA)

Dihydroxyphenylacetic acid
(DOPAC)

Ascorbic acid (AA)

Dihydroxybenzylamine
(DHBA)

Dicyanohydroquinone
(DCHQ)

FIG. 29.   Structures of several redox systems used to evaluate
carbon electrode performance.

Superficially DA and other hydroquinones appear to undergo a quasi-reversible 2e⁻ oxidation at low pH, with the stoichiometry of Eq. (19) [108,109].

$$QH_2 \rightleftarrows Q + 2e^- + 2H^+ \tag{19}$$

The apparent $E^{o'}$ shifts negative with pH by ca. 60 mV/pH unit for all catechols for the pH range 0–8 [50]. At a pH above about 4, DA undergoes a cyclization reaction to produce a di-hydroxyindole derivative [110], an overall 4e⁻ process. DOPAC, 4MC, and DHBA cannot cyclize, and in many studies of electrode kinetics the cyclization is too slow to observe for DA. Its effects will be negligible for most cases, and it will be ignored in further discussions.

Since 2e⁻ and 2H⁺ are involved in the oxidation of QH_2, the mechanism must involve several steps. Deakin and Wightman have examined the heterogeneous oxidation of catechols in detail [108,109] with two main conclusions. First, the process is characterized by two formal potentials, but only one oxidation peak is observed because the potential of the second electron is more negative than that of the first. For 4MC at pH 6.0, the first electron removed has a potential of 0.358 V vs. SCE, and the second, 0.045 V. The second conclusion involves the order of removal of e⁻ and H⁺ during the oxidation. For 4MC below pH 2, the sequence is e⁻ H⁺ e⁻ H⁺, meaning that 4MC oxidizes before it deprotonates. At pH 4, the most common route is e⁻ H⁺ H⁺ e⁻. These articles [108,109] also provided the microscopic rate constants for the protonation equilibria and electron transfers involved with the nine possible species in the QH_2/Q system. The apparent k^o values for the two oxidations of 4 MC at pH 6.0 on carbon paste are: k_1^o, 0.047 cm/sec; k_2^o, 0.028 cm/sec.

For the case of several catechols on a carbon paste electrode, the detail just discussed can be achieved. In general, however, a single quasi-reversible 2e⁻ reaction is observed, and ΔE_p is used to assess ET rates. If one desires a benchmark for ET kinetics, ΔE_p is usually sufficient, with faster k^o yielding smaller ΔE_p. However, details of the ET mechanism may depend strongly on the stepwise electron and proton transfer rates. Table 7 lists apparent $E^{o'}$ values for several catechols at pH 2.0 and 6.0. Conditions and techniques vary for the different compounds, and the values listed should be considered to be approximate.

c. Chemically Irreversible Organic Systems

Several irreversible organic redox reactions have been used to assess carbon electrode activity. In most cases the reaction mechanism is complex, with stepwise electron transfers or extensive reactions of the product, or both. The oxidation of ascorbic acid (AA) is particularly sensitive to carbon surface preparation, and it has been used frequently as a measure of electrode activity, with a more negative anodic peak potential implying a faster ET rate. The mechanistic details of the AA system are relevant to the current discussion and will be

TABLE 7

Experimental $E^{o'}$ Values for Several Hydroquinones
(All V vs. SCE)

	0.1 M H_2SO_4	pH 6	Ref.
Catechol	0.48		55
4-Methyl catechol		0.154	109
Dopamine	0.52	0.17	111
Dihydroxybenzyl amine		0.23[a]	111
Dihydroxyphenol acetic acid	0.50	0.20 / 0.17[a]	111 / 109
6-Hydroxydopamine	+0.27	−0.09	111
2,3-Dicyanohydroquinone	+0.69		55

[a]Adjusted to pH 6.0 assuming −60 mV/pH unit.

summarized below. The mechanisms of several other irreversible systems such as hydrazine oxidation, NADH oxidation, and phenol oxidation are beyond the scope of this chapter and will only be mentioned as they bear on the subject of carbon electrodes.

The overall oxidation of AA is shown in Fig. 30. Since the pK_a of AA is 4.2, a plot of E_p vs. pH exhibits a change in slope at pH 4.2. The hydration of the triketone is very fast [112], yielding a submillisecond lifetime for the dehydroascorbic acid (DHAA) oxidation product. The hydrated DHAA undergoes slow hydrolysis and degradation on a sufficiently long time scale to be irrelevant to discussions of ET kinetics. The main consequences of the hydration reaction is the inability to observe a reverse wave for AA oxidation, except at scan rates above 100 V/sec [112]. The ET kinetics of AA are very fast on Hg electrodes, and this fact enabled the determination of k_c. Vavrin obtained polarograms for AA from pH 2.2 to 8.3 [113].

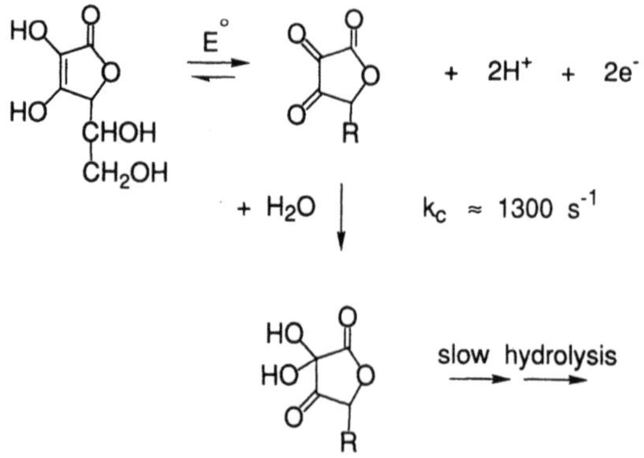

FIG. 30. Initial reactions in oxidation pathway of ascorbic acid. Slow hydrolysis results in ring opening and degradation on a time scale which is long compared to most electrochemical experiments.

Interpolation of his data provides an $E_{1/2}$ of -24 mV vs. SCE
at pH 7.0 and +237 mV at pH 2.0. These values are about 150
mV positive of the equilibrium potentials obtained from potentio-
metric titrations. The discrepancy is caused by the large equilibrium constant for hydration of the DHAA species, which reduces
the DHAA concentration during the titration, resulting in an observed $E^{o'}$ which reflects the AA/DHAA(H_2O) couple rather than
AA/DHAA. The fast hydration reaction causes a negative shift in
the polarographic $E_{1/2}$ [114], and voltammetric data obtained at
higher scan rates will be more reliable. At pH 7.4 and 0.1 V/sec,
E_p for AA on mercury is ca. +10 mV vs. SCE [115], and at pH 7.2,
200 V/sec, it is +12 mV [112]. Adjustment of these values to
pH 7.0 yields an approximate voltammetric E_p of 15–20 mV vs.
SCE on mercury.

Although ET kinetics for AA are very fast on mercury
electrodes, AA oxidation is much slower on most carbon electrodes. The mechanism of the AA oxidation has been the subject of several investigations [109,116-118]. For AA oxidation
on carbon paste, the observed E_p was independent of pH for
the range 4.8–7.8, where the reactant AA is deprotonated.
This implies that there are no protons involved in the rate-
limiting step, and that the first e^- is rate limiting on carbon
paste. Thus the scheme of Eqs. (20–22) was concluded [109],
with AA denoted as H_2AA to explicitly include protons.

$$HAA^- \rightleftarrows HAA\cdot + e^- \qquad\qquad (20)$$

$$HAA\cdot \rightleftarrows AA^{\cdot-} + H^+ \qquad\qquad (21)$$

$$AA^{\cdot-} \rightleftarrows DHAA + e^- \qquad\qquad (22)$$

On properly prepared GC electrodes the kinetics changed
significantly and E_p moved negative due to faster ET kinetics.
Two laboratories [109,118] independently arrived at the same
conclusion about the mechanism of HAA^- oxidation on activated

surfaces. At the beginning of the oxidation peak (below 0.0 V vs. SCE at pH 7.0), the second e$^-$ transfer is rate limiting, yielding an apparent transfer coefficient close to 1.5. At higher potential, the first e$^-$ transfer becomes rate limiting, and an apparent transfer coefficient of 0.5 is observed.

Although the oxidation mechanisms of both AA and catechols are complex, they have become useful benchmarks for ET activity of different carbon electrodes. We will see that the ET rates for $Fe(CN)_6^{-3/-4}$, AA, and dopamine are strongly dependent on electrode surface condition and provide useful insights into structural variables affecting the ET activity of carbon surfaces.

d. Effects of Surface Heterogeneity on Voltammetry

In the discussion of carbon surface variables, we considered the possibility that a nominally flat carbon surface has edge and basal plane domains, which differ in their properties. In the case of ET kinetics, edge and basal regions may have different k^o values. Furthermore, the edge and basal regions may vary greatly in size and spacing, and small active sites may produce radial rather than planar diffusion of electroactive species. The problem of kinetic heterogeneity has been considered by several authors [119–121] and is relevant to the discussion below.

Consider a flat electrode consisting of active regions of radius R_H, having some rate constant k_H^o for a particular redox system, and a total fractional coverage of the electrode surface of f_H. Furthermore suppose the remainder of the electrode has a rate constant k_L^o, and fractional coverage $1 - f_H$. Finally, assume the active regions are separated by a distance of $2R_L$, and that $k_H^o \gg k_L^o$. The magnitude of R_H and R_L relative to $\sqrt{Dt}$ can have a significant effect on the voltammetry observed with the heterogeneous surface [119]. There are three cases of interest here, depicted qualitatively in Figs. 31 and 32. For case 1, the electrode regions are small and closely spaced relative to the diffusion layer thickness, i.e., $R_H, R_L \ll \sqrt{Dt}$. In

Case 1: $R_H, R_L \ll \sqrt{Dt}$

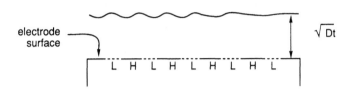

electrode
surface

$\sqrt{Dt}$

L H L H L H L H L

Case 2: $R_H < \sqrt{Dt} < R_L$

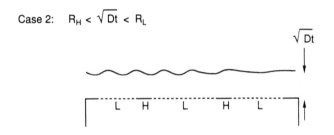

$\sqrt{Dt}$

L H L H L

Case 3: $R_H, R_L \gg \sqrt{Dt}$

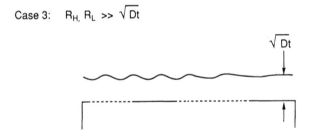

$\sqrt{Dt}$

FIG. 31. Diffusion regimes for a collection of low and high activity electron transfer sites on carbon electrodes. R_H denotes the mean radius of high activity sites with rate constant k_H^o; R_L and k_L^o denote the same parameters for low activity sites. Wavy line denotes approximate diffusion layer boundary.

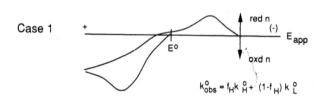

Case 1

$k_{obs}^o = f_H k_H^o + (1 - f_H) k_L^o$

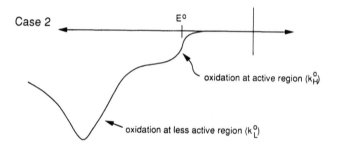

Case 2

oxidation at active region (k_H^o)

oxidation at less active region (k_L^o)

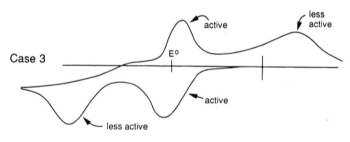

Case 3

FIG. 32. Voltammograms for a quasi-reversible system corresponding to cases of Fig. 31. f_H indicates fractional area of
high activity regions.

this case, a diffusing molecule can react at either region, and
the observed rate constant will be a weighted average of that
for each region, as in Eq. (23).

$$k^o_{obs} = f_H \, k^o_H + (1 - f_H) \, k^o_L \qquad R_H, R_L \ll \sqrt{Dt} \qquad (23)$$

For case 2 the active regions are widely separated and are small
compared to $\sqrt{Dt}$, i.e., $R_L < \sqrt{Dt} < R_L$. Each active region acts
as a microelectrode with convergent diffusion and an apparent
potential near $E^{o'}$ if k^o is large. The much larger, less active
region behaves as a planar electrode with low k^o_L, yielding a
normal peak shaped wave (Fig. 32). In case 3, the active regions are large and widely spaced compared to $\sqrt{Dt}$, i.e.,
$R_H, R_L \gg \sqrt{Dt}$. In this case, the electrode regions are diffu-
sionally decoupled and will behave like two physically separated
electrodes with two different k^o values. Two independent voltammetric waves will occur with ΔE_p values reflecting the differences in k^o.

3. Adsorption

In addition to observations of capacitance and electron transfer
rates, adsorption of solution components on carbon electrodes
is also used to characterize the surface. As noted in Sec.
III.A.5, an adsorbed electroactive species is easily detected
and quantified by voltammetry or chronocoulometry, with detection limits of 0.01 monolayer or lower. For example, consider
a strongly adsorbed electroactive molecule like phenanthrenequinone (PQ). PQ is very slightly soluble in 1 M $HClO_4$ (ca.
10^{-6} M), and a voltammetric peak for solution PQ is too small
to detect. The voltammetry of adsorbed PQ is easily detected
if sufficient time is allowed to preconcentrate the dilute solution
PQ on the electrode (ca. 1 hr). The charge, Q, determined
by integrating the voltammetric peak is simply related to the
moles adsorbed [57].

The PQ case is simple because the adsorbing molecule in solution yields a negligible current, and one monitors only adsorbed PQ. It is more common that a solution species yields currents from both adsorbed and diffusing material. The occurrence of adsorption may be obvious, with a separate voltammetric wave or a linear dependence of i_p with a scan rate. In many cases, however, the adsorption is less obvious and careful quantitative techniques, particularly chronocoulometry, are required [96]. Semi-integral voltammetry has also proved useful for detecting adsorption on carbon electrodes [122]. If the semi-integrated current vs. potential plot is peak shaped rather than sigmoidal, adsorption of the reactant is indicated.

IV. CARBON ELECTRODE PREPARATION AND PERFORMANCE

The previous two sections discussed the many variables which determine carbon surface structure and ways to characterize the electrode surface. Given the overall goal of relating surface characteristics to electrochemical performance, the question arises of how carbon type and preparation affect the state of the carbon surface in electrolyte solution. In order to determine how surface variables affect ET rates or background current, a reproducible means to prepare the electrode surface is essential. Thus, the first requirement for relating surface structure to electrochemical performance is a reproducible pretreatment procedure. The second requirement is a means to determine the surface structure and composition. If these two prerequisites are met, the dependence of ET rate and background current on structure can be addressed, since a known surface will be available.

Unfortunately, well-defined carbon surfaces are difficult to prepare, and one rarely knows the surface structure when conducting electrochemical experiments. A staggering number of

pretreatment procedures have been developed for carbon electrodes, and there is currently no consensus on how to prepare
a reproducible surface. However, a variety of results from
many different laboratories have identified several critical surface variables which affect electrode performance. In this section, we will describe pretreatment procedures for several different forms of carbon, and compare the electrochemical characteristics of the resulting carbon surface. Of particular interest
will be the background current and electron transfer rates for
benchmark redox systems. The discussion of pretreatment
procedures is not intended to be comprehensive, but rather to
identify the critical surface variables. Pretreatment procedures
and electrochemical performance will be discussed for each type
of carbon. Carbon materials will be considered in order of increasing microstructural complexity, starting with the best defined carbon type, HOPG. This order was chosen for clarity
and is not chronological, nor does it follow any ranking of performance. After discussing the many pretreatments and their
resulting electrochemical behavior, some conclusions and hypotheses about structure/reactivity correlations will be considered
in Sec. V.

A. Highly Ordered Pyrolytic Graphite

Although HOPG is not as commonly used as several other carbon
materials, several useful investigations of its electrochemical
properties have occurred and are discussed below. Its well-
defined structure plus the ability to prepare basal plane HOPG
surfaces with low defect concentration make it attractive for relating structure to reactivity. Three preparation methods will
be discussed: cleavage and exposure of basal plane, laser activation, and electrochemical activation.

1. Cleavage of HOPG

Since the interplanar bonding in HOPG is very weak, the layers
are easily separated mechanically. The common procedure involves pressing ordinary "Scotch" tape on the basal plane surface, then removing the tape and attached graphite layers. As
shown by STM, the basal surface thus exposed is predominantly
atomically flat, with relatively low defect density. The exposed
surface is fragile and is easily damaged by mechanical manipulation. As noted below, surface defects have a major effect on
observed ET kinetics, so great care is required to minimize surface damage [92]. To conduct electrochemical experiments on
the basal plane, an electrode may be defined by an o-ring held
in place by a teflon washer with gentle pressure. The voltammetry of $Fe(CN)_6^{-3/-4}$ on basal plane HOPG is shown in Fig.
33. The electron transfer rate is very slow, with $\Delta E_p > 700$
mV, and similar behavior is observed for dopamine [31,92,115].
If damage occurs to the basal plane surface, either ΔE_p for

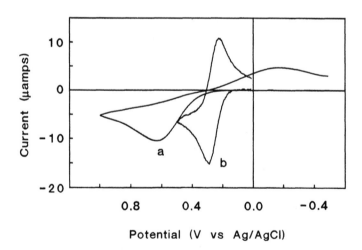

FIG. 33. Voltammograms of $Fe(CN)_6^{-3/-4}$ (1 M KCl) on basal
plane HOPG (curve a) and polished edge plane HOPG (curve
b). $\nu = 0.2$ V/sec.

$Fe(CN)_6^{-3/-4}$ will be less than 700 mV or, in severe cases, two peaks will be observed due to fast ET on the damaged region. Thus slow charge transfer for $Fe(CN)_6^{-3/-4}$ is a reliable criterion for surface quality. The $k°$ for $Fe(CN)_6^{-3/-4}$ estimated from ΔE_p for basal plane is below 10^{-7} cm/sec [31,115].

The capacitance of basal plane HOPG is anomalously low, with values of 1.9 $\mu F/cm^2$ (92) and 3.0 $\mu F/cm^2$ [42,43,123] being reported. This low capacitance has been attributed to a space charge layer caused by the semimetal character of HOPG [4,42, 43,123]. As shown in Fig. 26, curve a, the basal plane behaves as a nearly ideal capacitor, with no observed frequency dispersion.

The edge plane of HOPG may also be examined electrochemically, but the surface is poorly defined. The HOPG piece may be mounted in epoxy (e.g., Torrseal, Varian) and polished. Because of the possibility that small ragged edges of HOPG may fold over the edge plane, it is not likely that the exposed surface is pure edge plane. Nevertheless, the edge plane surface has dramatically different electrochemical characteristics compared to basal plane. The ET rate is much higher, with ΔE_p for $Fe(CN)_6^{-3/-4}$ of about 70 mV at 0.2 V/sec (Fig. 33). The $k°$ for edge plane varies for different trials, but is about 0.06– 0.1 cm/sec, or at least 10^5 times the basal plane value [92]. The observed capacitance of edge plane varies as well, but is very high, at least 60 $\mu F/cm^2$ [92,124]. Given the uncertainty of edge plane structure, several phenomena may contribute to the observed capacitance, but it is clearly 20–30 times the basal plane value.

The main conclusion from comparing edge and basal plane HOPG is the extreme anisotropy of $k°$ and $C°$. The basal surface is an extremely unreactive surface for electron transfer to $Fe(CN)_6^{-3/-4}$ or DA in water, and the observed capacitance is only 5% that of edge plane. Clearly the distribution of basal

and edge plane will be important to the electrochemical behavior
of any carbon surface. Since the ET rate anisotropy of HOPG
was revealed quite recently [31,92,115], it is not yet known if
the effect is observed in solvents other than water or redox
systems other than $Fe(CN)_6^{-3/-4}$, DA, or AA. The observation
of extreme rate anisotropy also explains the need for care when
studying the basal plane. Since edge plane has at least 10^5
times higher k°, even small surface defects on basal plane HOPG
will introduce enough edge plane to significantly increase the
observed k° over the value for pure basal plane. In fact, the
observed anisotropy of about 10^5 may actually be much larger
if a perfect basal plane surface could be prepared.

2. Laser Activation of HOPG

Of the many methods for improving ET rates on carbon electrodes,
laser activation is a recent example. Reported initially in 1984
[125] for GC and Pt electrodes, laser activation improves k°_{obs}
for AA, DA, and $Fe(CN)_6^{-3/-4}$ by means of short (ca. 10 nsec)
intense (ca. 30 mW/cm^2) Nd:YAG laser pulses delivered to the
electrode surface in situ [126,128]. Limiting the current discussion to HOPG, it was found that a 50 MW/cm^2 laser pulse
increased the basal plane k° for $Fe(CN)_6^{-3/-4}$ by a factor of
10^4–10^5 [31,92]. Figure 34 shows voltammograms of $Fe(CN)_6^{-3/-4}$
on basal plane HOPG before and after laser pulses of 50 and 80
mW/cm^{-2}. Note the large decrease in ΔE_p after the 50 MW/cm^2
laser pulse, but only a slight further decrease when an 80 MW/
cm^2 pulse is employed. The plot of k° vs. laser power density
in Fig. 35 shows that the activation of the HOPG basal plane
has a well-defined threshold at 45 MW/cm^2, but at higher power
densities the observed k° is erratic [92].

The large increase in rate after laser activation is accompanied by changes in the Raman spectrum and capacitance. Figure
26 shows the appearance of the 1360 cm^{-1} Raman band following
the laser pulse, with severe carbon disorder apparent at 90

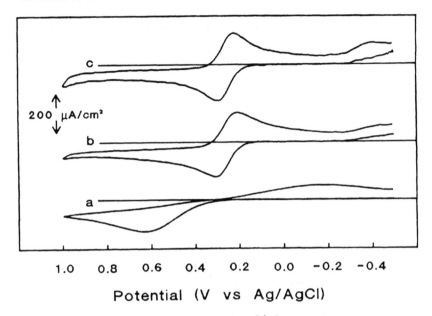

FIG. 34. Voltammograms of $Fe(CN)_6^{-3/-4}$ on basal plane HOPG in 1 M KCl (ν = 0.2 V/sec). Curve a is initial, b and c are after three 50 MW/cm^2 and three 90 MW/cm^2 laser pulses, respectively. (From Ref. 92, reprinted with permission.)

MW/cm^2. Clearly 50 MW/cm^2 laser pulses decrease L_a and create edge plane on the surface. The capacitance measured with a 100 Hz triangle wave increases from 1.9 μF/cm^2 on fresh basal plane to about 15 μF/cm^2 after laser activation at 90 MW/cm [92]. Although C_{obs}^o increases by a factor of 8, recall that k^o increases by a factor of 10^5 over the same power density range. C_{obs}^o and k_{obs}^o both exhibit a threshold in power density of 45 MW/cm^2. The clear qualitative conclusion from these experiments is that a laser pulse delivered to HOPG creates edge plane, and since k^o on edge plane is much higher than basal plane, a large increase in k^o is observed following laser activation.

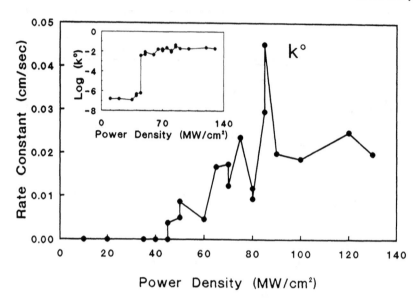

FIG. 35. Observed k° on basal plane HOPG following three laser pulses at various power densities. Fresh basal plane was exposed before each measurement. Inset is same data on log scale. (From Ref. 92, reprinted with permission.)

The conclusion that fast electron transfer occurs at edge plane defects is reinforced by an experiment involving microscopy of electrogenerated chemiluminescence (ECL) developed by Engstrom [129]. HOPG basal plane in a solution of luminol and H_2O_2 was examined with high sensitivity fluorescence microscopy. When an oxidizing potential was applied, ECL occurred at surface sites where the luminol was oxidized. Intentional defects on the basal plane exhibited much brighter ECL than the undamaged basal plane, indicating that faster electron transfer occurred where edge plane was present.

The nearly ideal shape of the voltammogram b of Fig. 34 indicates that the laser-induced edge sites and spacing on HOPG are small and closely spaced, i.e., that case 1 from Figs. 31

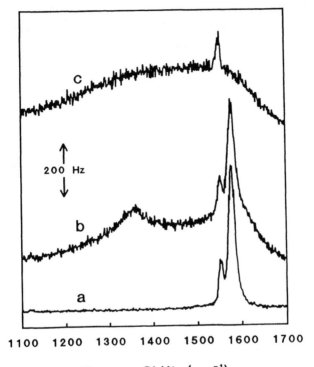

FIG. 36.   Raman spectra of basal plane HOPG following laser
activation.   Spectrum a is initial, b and c are after 50 MW/cm^2
and 90 MW/cm^2 activation, respectively.   Peak at ca. 1560 cm^{-1}
is atmospheric oxygen.   (From Ref. 92, reprinted with permission.)

and 32 applies.   The amount of edge plane generated by the
laser (f_e) can be estimated from the changes in k^o_{obs} and C^o_{obs}.
If k^o_{obs} and C^o_{obs} are weighted averages of edge and basal plane
contributions, Eqs. (24) and (25) should be valid.

$$k^o_{obs} = k^o_e f_e + k^o_b (1 - f_e) \qquad (24)$$

$$C^o_{obs} = C^o_e f_e + C^o_b (1 - f_e) \qquad (25)$$

From independent measurements on edge and basal plane
HOPG, $C_b^o = 2 \ \mu F/cm^2$, $C_e^o \cong 70 \ \mu F/cm^2$, $k_b^o < 10^{-7}$ cm/sec, and
$k_e^o \approx 0.1$ cm/sec [92]. Combining these values with Eqs. (24)
and (25), f_e may be estimated from either C_{obs}^o or k_{obs}^o, with
the results shown in Fig. 37. This approach requires assumptions that roughness factor and surface cleanliness do not change
significantly with laser activation and that edge and basal plane
capacitance and rate constants are additive. Although the f_e
vs. power density plots in Fig. 37 show significant scatter; it
is clear that C_{obs}^o and k_{obs}^o lead to similar estimates of f_e, even
though k_{obs}^o changes by a factor of 10^5 while C_{obs}^o changes by
only a factor of 8. Equation (24) reinforces the statement that
basal plane defects can drastically affect k_{obs}^o on "fresh" basal
plane. For example, 1% edge plane on a basal surface will yield

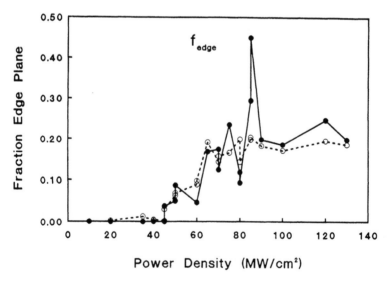

FIG. 37. Fraction edge plane, f_e, calculated from k_{obs}^o (solid
line) and C_{obs}^o (dashed line). (From Ref. 92, reprinted with
permission.)

a k^o of 10^{-3}, at least four orders of magnitude higher than the
"pure" basal plane k^o_{obs}. Finally, if there are sites on the carbon surface, the results in Figs. 34–37 imply that these sites
exist on graphitic edge plane.

SEM revealed little about the morphology of the laser-
irradiated HOPG, because the laser-induced damage appeared to
be too small to resolve [31]. However, Dresselhaus et al. have
studied the morphology of laser-irradiated HOPG during attempts
to make liquid carbon [130–132]. They found numerous upheavals of graphite planes in the irradiated region, as well as sphe-
roids which appeared to result from solidification of liquid carbon.
The upheavals formed a network of exposed edge plane on the
surface, but no quantitative estimate of f_e was made. They
propose a mechanism for surface damage involving constrained
thermal expansion in the irradiated region, causing the lattice
to shatter and, at high powers, liquefy [132]. The laser
causes the graphite to rapidly expand in both the c- and a-
direction, while the surrounding cool graphite constrains the
expansion. This damage mechanism is most effective for short,
intense laser pulses.

In summary, laser activation of HOPG has reinforced the
conclusion from Sec. IV.A.1: There must be graphite edge
plane present to have rapid electron transfer. Furthermore,
the observed k^o and C^o on laser-damaged basal plane are consistent with a weighted average of the corresponding edge and
basal plane values.

3. Electrochemical Pretreatment (ECP) of Basal Plane HOPG

ECP of carbon has been a major research subject because it can
greatly increase ET rates for a variety of redox systems. The
vast majority of the ECP research has been on GC and carbon
fibers and will be discussed later. However, Wightman et al.
applied ECP to HOPG as early as 1978 [133], and its effects on

HOPG are quite instructive. ECP involves application of one of a variety of potential waveforms to a carbon electrode in solution. In the case of HOPG, these include a constant E_{app} of 1.4–1.5 V vs. SCE in citrate buffer [115,133], a fixed potential of 1.8 V vs. SCE in 0.1 M KNO_3 [31], or cyclic potential scans from 0.24 to 2.04 V vs. SCE at 200 mV/sec [78]. The ECP treatment is usually applied for several minutes and may involve a reduction step at −0.1 V vs. SCE [31,134]. It has been demonstrated that ECP of carbon generates an oxide film [5] but such a film has not been explicitly confirmed for HOPG.

ECP of HOPG dramatically affects its ET activity, as shown in Fig. 38 [31]. Both DA and $Fe(CN)_6^{-3/-4}$ exhibit ΔE_p greater than 700 mV on untreated basal plane (Fig. 33). ECP (1.95 V for 2 min, −0.1 V, 30 sec) reduces ΔE_p to about 100 mV for $Fe(CN)_6^{-3/-4}$ and 200 mV for DA. Similar large improvements in ET rates were observed for AA [115] and DA [133] for ECP in citrate buffer (20 min at 1.4 V). These results indicate an increase in $k°$ for $Fe(CN)_6^{-3/-4}$ from $<10^{-7}$ to 0.001–0.006 cm/ sec, or about 4 orders of magnitude. As with laser activation, the appearance of the voltammograms in Fig. 38B indicates that any kinetic heterogeneity on the surface is small compared to $\sqrt{Dt}$ (case 1 of Figs. 31 and 32).

Accompanying these electrochemical improvements induced by ECP are some major morphological and spectroscopic changes. At the SEM level, surface defects are visible [31,133], sometimes several tens of micrometers in size. At the STM level, it is possible to observe the formation of "ridges and valleys" (Fig. 25). The morphological results support a nucleation and growth mechanism [31,78] in which oxidation of the graphite begins at a defect (probably some exposed edge plane), then progresses along the edge plane, forming long, large regions of oxidized edge plane. The Raman spectroscopy of HOPG before and after ECP at two different oxidation potentials is shown in Fig. 39.

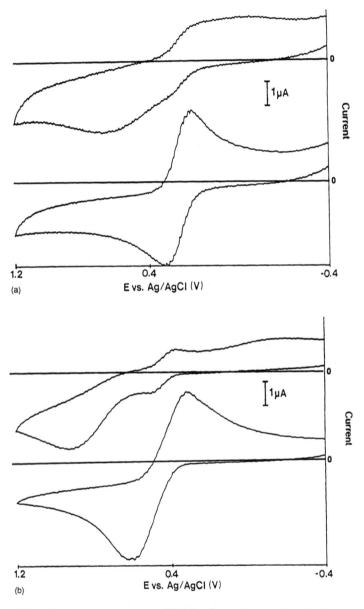

FIG. 38. Voltammograms on HOPG after electrochemical pretreatment for 2 minutes in 0.1 M KNO$_3$. Upper voltammogram in both sets is after 1.85 volt ECP, lower is after 1.95 volt ECP. Curves marked A are Fe(CN)$_6^{-3/-4}$, those marked B are dopamine. (From Ref. 31, reprinted with permission.)

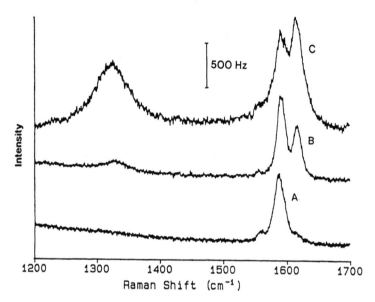

FIG. 39. Changes in Raman spectrum following ECP of HOPG at 1.6 V (A), 1.85 V (B) and 1.95 V (C). Laser spot covered ca. 0.1 mm^2 of the carbon surface. (From Ref. 31, reprinted with permission.)

As the surface is oxidized, a 1620 cm^{-1} band appears, then later a 1360 cm^{-1} band. As with laser activation, appearance of the 1360 cm^{-1} band coincides with the decrease in ΔE_p for Fe(CN)$_6^{-3/-4}$ or DA [31]. The results indicate that the HOPG initially delaminates, then fractures, probably because of lattice strain induced by the oxidation process [31,135].

When ECP is carried out under milder conditions (i.e., lower anodization potential), both the voltammetry and Raman spectroscopy are altered. Figure 38 shows a DA voltammogram after mild ECP at 1.85 V. Note that two redox couples occur, one with small ΔE_p (high k^o) and one with large ΔE_p (low k^o). This behavior is expected for case 3 (Figs. 31–32), in which the surface heterogeneity is large relative to $\sqrt{Dt}$. A Raman

microprobe which samples only a 5-μm spot on the surface confirmed the heterogeneity. As shown in Fig. 40, spectra taken
20 μm apart exhibit greatly different 1360 cm^{-1} intensity, implying that surface damage is not uniform across the surface.
Since $\sqrt{Dt}$ for the voltammetry was about 20 μm, the spatial
heterogeneity on the surface treated at 1.85 V is at least this
large [31].

The observed k° for $Fe(CN)_6^{-3/-4}$ for both laser and ECP
surfaces correlated with the 1360 cm^{-1} band intensity. For example, a laser-treated surface with a 1360/1582 Raman intensity
ratio of 0.2 had a k°_{obs} of 0.0018 cm/sec, while an ECP surface
with a ratio of 0.6 had a k°_{obs} of 0.0063 cm/sec [31]. Despite

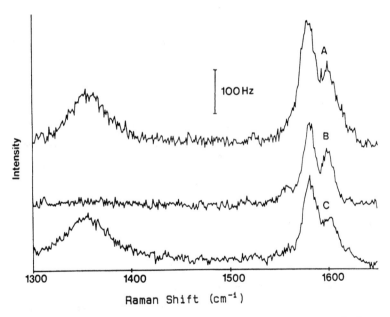

FIG. 40. Raman microprobe (5 μm spot size) spectra of the
1.85 volt ECP surface from Fig. 39. Microprobe was translated
ca. 20 μm between each spectrum. (From Ref. 31, reprinted
with permission.)

the very different mechanisms of the two activation procedures, the spectroscopically determined degree of carbon disorder correlated with ET kinetics. Not only is it crucial to have edge plane to observe fast ET rates, but it does not appear to matter how edge plane is produced. Furthermore, the two procedures should result in very different degrees of surface oxidation, yet they yield comparable rates. For ECP, oxidation provides a means to cause lattice damage, but the important thing appears to be the formation of edge plane, not surface oxides per se [135].

On a brief note of electroanalytical significance, the low capacitance of basal plane HOPG makes it attractive as an amperometric detector. Provided it is activated, the background current is low, and a wide potential range is available. ECP-activated HOPG was initially evaluated as an amperometric LC detector [133] as an alternative to carbon paste or Hg electrodes.

B. Pyrolytic Graphite

As noted earlier, pyrolytic graphite (PG) is an ordered graphite with a smaller L_a and L_c than HOPG. A cleaved PG basal surface is not as flat or smooth as HOPG, and conical islands are visible to the eye. Although PG has been used in many electrochemical projects, it currently is much less popular than GC or carbon paste.

1. Cleaved PG

Panzer and Elving [136,137] concluded that the cleaved basal surface of PG is much more reproducible than polished material and has a much lower background current. Based on the Raman spectrum of Fig. 14, cleaved basal plane PG has a smaller L_a than HOPG and therefore more exposed edges. From the $Fe(CH)_6^{-3/-4}$ results on edge vs. basal HOPG, we would predict that cleaved, basal PG should exhibit a k° that is much higher

than 10^{-7} cm/sec. Panzer and Elving report a ΔE_p for
$Fe(CN)_6^{-3/-4}$ of 130 mV for cleaved PG and 90 mV for ground
PG in 0.5 M KCl, pH = 8.5 and ν = 0.1 V/sec. These values
correspond to a k_{obs}^o of 0.007 cm/sec for ground PG and 0.002
for cleaved PG [138]. Even taking the difference in KCl concentration into account (a factor of 2) these values are much
higher than basal plane HOPG (by about 10^4) but lower than
HOPG edge plane. Although there are significant uncontrolled
variables involving cleanliness and roughness factor, cleaved
PG exhibits a k^o value that is consistent with Eq. (24). Since
the basal PG surface has much smaller L_a, and therefore much
higher f_e than HOPG basal plane, the $Fe(CN)_6^{-3/-4}$ rate is
correspondingly higher. At least for $Fe(CN)_6^{-3/-4}$, cleaved PG
behaves like damaged HOPG.

2. Polished Basal Plane PG or Edge Plane PG

As noted by Panzer and Elving, abrasion of PG basal plane
causes a large increase in background current which was attributed to microcracking and greatly increased microscopic surface area. Edge plane PG is prepared by mounting PG in epoxy
with the edge plane exposed, followed by polishing or other
treatment [8]. Since edge plane PG is more commonly used and
abraded basal plane PG will behave similarly, only edge plane
PG will be discussed in any detail. Polished edge plane PG
(EPG) exhibits a ΔE_p for $Fe(CN)_6^{-3/-4}$ of 75 mV, corresponding
to a k_{obs}^o of about 0.005 cm/sec. The fact that this value is
lower than edge plane HOPG is attributable to differences in
surface preparation. As shown in Table 8, reaction of the EP
surface with a silanizing reagent had little effect on ET reversibility for $Fe(CN)_6^{-3/-4}$, $Ru(NH_3)_6^{+2/+3}$ or a ferrocene derivative
[139], consistent with the earlier findings of Elliot and Murray
[140]. Thus EPG is a reasonably active surface for ET, as expected from the HOPG results.

TABLE 8

ΔE_p (ν = 20 mV/sec) for Various Redox Couples on
Native and Trimethylsilane Modified Edge
Plane PG Electrodes

	ΔE_p, mV native edge	ΔE_p, TMS modified[a]
$Fe(CN)_6^{-3/-4}$	75	75
$Ru(NH_3)_6^{+2/+3}$	60	60
Ferrocene monocarboxylic acid	60	60
Cytochrome c	65–70	125–130
Azurin[b]	95–100	100–105

[a]Trimethyl silyl groups attached to surface oxides.
[b]A "blue" copper protein.
Source: From Ref. 139.

Armstrong and Brown [139] pointed out (Table 8) that cyto-
chrome c (cyt c) was made less reversible when EPG was silanized. They attribute this decrease to the blockage of surface
alcohol or carboxylate groups to which cyt c binds during electron transfer. ΔE_p for cyt c was restored to 65–70 mV immediately after the silyl groups were removed by hydrolysis in base.
Armstrong and Brown conclude that surface oxygen sites accelerate cyt c ET by providing a binding site. Thus cyt c differs
fundamentally from $Fe(CN)_6^{-3/-4}$ because its ET is apparently
catalyzed by surface oxide.

These observations were extended in two subsequent papers
[141,142]. It was found that $Cr(NH_3)_6^{+3}$ inhibited cyt c ET by
binding to surface oxide groups. Plastocyanin (Pc), an anionic
redox protein, behaved very differently from cyt c, however.
It exhibited fast electron transfer at EPG at pH 4, but slow ET
at pH 5. The authors attributed the pH sensitivity to deprotonation of surface carboxylates. Furthermore, $Cr(NH_3)_6^{+3}$ catalyzed
Pc electron transfer by occupying anionic surface groups. Both

the cyt c and Pc results are consistent with the involvement of anionic surface oxides in electron transfer, acting as catalysts for cyt c and inhibitors for Pc. Armstrong, et al. also examined the shape of the Pc and cyt c voltammograms, and they noted that under conditions where the ET is fast, the cyclic voltammograms had the expected peak shapes. However, under conditions where ET appeared slow, the waves were sigmoidal. Theory [141] and experiment [142] support the hypothesis that ET between a redox protein and a surface oxide is fast, but that the oxide site density can change significantly with conditions. For example, a high density of anionic surface oxide sites exhibit high k^o. As the number of active sites decreases, the wave shape becomes sigmoidal (case 2, Fig. 32). Thus the ET of cyt c to silanized EPG appears slow on a macroscopic scale, but is fast at a microscopic level to the few remaining oxide sites. Stated differently, a reported k^o is usually based on geometric area, but the local k^o at microscopic oxide sites may be much higher.

To summarize the results on PG, all observations are consistent with the conclusions reached with HOPG, except for the behavior of cyt c and PC. These cases are strongly affected by specific interactions with surface oxides, but no evidence indicated that similar interactions occur for $Fe(CN)_6^{-3/-4}$, $Ru(NH_3)_6^{+2}$, or ferrocene monocarboxylic acid.

C. Carbon Paste, Graphite Rods, Carbon Composites

With regard to carbon structure and characteristics, these electrode types are similar. They all involve powdered or randomly oriented graphite, usually combined with an insulating binder or filler. The carbon structure is similar to PG, except with smaller L_a and L_c and no ordering of the microcrystallite orientation. Carbon paste is the most commonly used of the three, due in part to its remarkably low background current and ease of surface

preparation. Although there is a rich literature on carbon
paste and other carbon composites, only topics related to electron
transfer at the carbon surface will be discussed here, and chemically modified carbon paste and composites will not be discussed.

1.   Carbon Paste (CP)

Electrodes made from mixtures of graphite and a low volatility
organic liquid were introduced by Adams in 1958 [143]. Not
only were they suitable for a variety of organic oxidations, but
they also had an extremely low background current. The low
background was a major advantage over Pt and Au electrodes,
which have large backgrounds in water from oxide formation and
reduction. The graphite used was Acheson grade 38, whose
modern equivalent is GP-38 from Ultracarbon. The development
of carbon paste led to widespread use of voltammetry for studying organic reaction mechanisms [144], and later the development
of electrochemical detectors for liquid chromatography [145].
An additional major advantage of carbon paste is the ease of
renewal of the surface, providing a fresh surface unaffected by
electrode history. In keeping with the scope of this chapter,
only the electrode kinetics and background characteristics of
carbon paste will be discussed.

The composition of CP can affect background current, ET
kinetics, and voltammetric peak height. Rice, Galus, and Adams
[146] varied the identity and weight fraction of the pasting liquid and determined k^o_{obs} for $Fe(CN)_6^{-3/-4}$ in 1 M KCl. For 2/1
(w/w) mixtures of graphite and organic liquid, k^o_{obs} ranged
from 3.1×10^{-3} cm/sec for hexane to 0.11×10^{-3} cm/sec for
Nujol. The decrease in k^o_{obs} for heavier hydrocarbons was
attributed to the difficulty of removal of organic liquid from the
graphite surface. It was also noted that anodization accelerated
k^o, apparently by making the graphite more hydrophilic and less
prone to hydrocarbon adsorption. Chemical oxidation of the
graphite before pasting also increased the resulting k^o. Although

a rate constant was not reported, the authors state that unpasted graphite powder showed the highest k^o for $Fe(CN)_6^{-3/-4}$, close to that of Pt. They conclude that the pasting liquid inhibits ET, and note that hexadecane provided the best trade-off of ET kinetics and ease of use [146].

Urbaniczky and Lundstrom [147] varied the type of graphite rather than the pasting liquid, and noted large effects of carbon type and history on k^o. They noted that k^o increased and C_{obs}^o decreased for all graphites studied if the carbon was heat treated before pasting. For example, heat-treated Ultracarbon UCP-1-M in hexadecane resulted in $k_{obs}^o = 1.5 \times 10^{-3}$ cm/sec and $C_{obs}^o = 2.1$ $\mu F/cm^2$. It was concluded that surface oxides inhibit ET to $Fe(CN)_6^{-3/-4}$ and cause a high C_{obs}^o due to surface redox reactions.

The low background current of CP arises from the relatively small fraction of the surface that is exposed carbon. Small, reactive graphite particles protruding from a field of inert hydrocarbon will show very low observed capacitance since so little of the electrode surface is carbon. Since the electroactive species can undergo radial diffusion to the graphite particles, a higher faradaic current than expected from the microscopic area results. Thus the background is determined by the active graphite area, while the faradaic current is proportional to the entire electrode area [148], as in Figs. 31–32, case 1.

The low observed capacitance reported for CP [147] of ca. 2 $\mu F/cm^2$ is not the capacitance of the carbon, which should be 20–70 $\mu F/cm^2$, but rather the apparent capacitance of the carbon/ binder composite. Thus, the organic liquid in CP greatly decreases the background current by sealing pores and reducing the active carbon area, but it inhibits k^o by adsorption on the carbon and decreasing the active area. In many applications, e.g., LC/EC, the gain in reduced background is worth the cost in ET kinetics. However, any systematic attempts to use

CP to investigate ET kinetics will probably be frustrated by interference of the pasting liquid.

2. Graphite Rods

Graphite rods are prepared as described in Sec. II and are
randomly oriented graphite with L_a and L_c on the order of a
few 100 Å. The electrochemical literature usually refers to such
materials as spectroscopic graphite, since it is used in atomic
spectroscopy and has a very low level of metal impurities. If
graphite rods are used as received, the background current is
usually unacceptably high due to penetration of electrolyte into
the pores of the rod, and the resulting high surface area exposed to the solution. When the graphite part is impregnated
with a water-immiscible material, an electrode similar to carbon
paste results. A common technique is impregnation by heated
paraffin or ceresin wax in a vacuum to make a wax-impregnated
graphic electrode (WIGE). WIGEs are relatively easily resur-
faced and share with CP a low background current, since the
wax prevents permeation of the graphite pores by electrolyte,
and the exposed active surface area is small. Several alternative impregnation procedures and electrode compositions have
been described [2]. WIGEs have been used for trace metal analysis, particularly as a substrate for Hg film deposition before
anodic stripping voltammetry. The author is not aware of any
systematic studies of electrode kinetics on the WIGE.

3. Carbon Composites

A drawback of carbon paste electrodes is their limited utility in
most organic solvents, due to dissolution of the paste or permeation of the solvent into the bulk of the electrodes. Partly for
this reason, a variety of alternative composites have been developed. These include polyethylene/carbon black [149], Kel-F/
graphite ("kelgraf") [148,150,151], carbon black immobilized in

cross-linked polystyrene [152,153], and epoxy/graphite [154,155].
Like carbon paste, the surfaces of these composite electrodes
are only partially carbon, so the active area and background
current are reduced. In the case of kelgraf, a detailed study
of sensitivity and S/N for its use as an LC detector revealed
significant improvements over glassy carbon [148,150]. Furthermore, the electrode is compatible with most solvents and is both
rugged and polishable.

Fast electrode kinetics was not a major objective of the development of composite electrodes, and little quantitative information is available. An advantage of composites over solid carbon electrodes is the ability to incorporate redox agents or
other modifiers in the binding materials [152,153]. By this
route, the catalytic or adsorption properties may be made quite
different from the native carbon and binder. For example, a
carbon black/polyvinylpyridine/polystyrene composite adsorbed
metal ions from solution and exhibited ion exchange to the immobilized pyridine [152]. As with carbon paste, the ability to
easily modify the chemical characteristics of the electrode permits a wide variety of useful applications.

In the context of electrode kinetics and background processes,
carbon paste—impregnated graphite rods and carbon composites
share some common traits. First, the background current can
be very low when the fraction of the electrode surface that is
carbon is small. Second, electron transfer performance is dependent on the characteristics of the carbon used, with inhibi-
tion by the organic binder quite likely. In order to understand
carbon structural effects on electrode kinetics, variables resulting from the organic binder are best eliminated. As we will see
in the next section, glassy carbon has been the subject of many
more kinetics investigations than carbon paste or composites.

D. Glassy Carbon

Unlike graphite, glassy carbon is not permeated by water or
organic solvents and does not require a binder to make a useful
electrode. In addition, it may be polished, exposed to high vacuum, heated and chemically derivatized, and is compatible with
organic solvents. Mainly for these reasons GC has been widely
used, and has superseded carbon paste and graphite electrodes
in many electroanalytical experiments. It is the most widely
studied carbon material in the context of electrode kinetics. As
one would expect, a wide variety of pretreatment procedures is
used on GC, and no single protocol has become a standard.
IUPAC has provided a compilation of $E^{o'}$ values for various redox
systems and anodic and cathodic limits for a variety of solvents
observed with GC, but no standard pretreatment procedures have
been proposed [7]. When particular pretreatment procedures are
described below, their effects on the k^o_{obs} and surface structure
will be noted.

1. Polishing

Nearly all pretreatments for GC start with polishing, so it will
be considered first. Since a GC piece received from a manufacturer has completely unknown surface properties, it is prudent
to remove some of the surface by polishing before conducting
electrochemical experiments. Most laboratories have taken the approach of adopting a uniform procedure to produce a surface with
unknown structure but reproducible properties. A large number
of polishing procedures have been devised, and they have been
compiled [83] and reviewed [6]. Representative procedures and
the resulting k^o_{obs} for $Fe(CN)_6^{-3/-4}$ are listed in Table 9. More
comprehensive discussions of polishing appear elsewhere [2,3].

Although polishing procedures are far from standardized,
several points are available from Table 9 and associated references. First, for $Fe(CN)_6^{-3/-4}$ k^o_{obs} varies from 10^{-4} to 0.14

TABLE 9

k^o for $Fe(CN)_6^{-3/-4}$ on Polished GC

GC type	Polishing procedure	Electrolyte	k^o (cm/sec)	Ref.
Tokai GC-20	Al_2O_3 on glass	1 M KCl	0.14 ± .01	83
Tokai GC-20	Al_2O_3 on glass	0.5 K_2SO_4	0.12 ± .02	83
Tokai GC-20	Al_2O_3 on polishing cloth	1 M KCl	0.002[a]	156
Tokai GC-10	Al_2O_3, metallographic	1 M KCl	0.07 ± .01	157
Normar GC	SiC, Al_2O_3, high speed		0.098 ± .045	158
GC-30	Al_2O_3, polishing cloth	1 M KCl	0.005 ± .003	159
GC-30	fractured, unpolished	1 M KCl	0.5 ± 0.2	159

[a]Calculated from ΔE_p given in Ref. 156.

cm/sec, depending on polishing procedure [83], indicating extreme sensitivity of k_{obs}^o to polishing history. Second, the use of a polishing cloth decreases k_{obs}^o significantly [83,134], probably by contamination of the carbon surface. If the final step involves Al_2O_3 on glass, high k^o can be observed. Third, impurities in the polishing materials are deleterious to fast ET kinetics. Scrupulous attention to cleanliness and water purity are a prerequisite to obtaining a high k^o [83]. Fourth, GC mounted in a holder (e.g., a Kel-F jacket) may be contaminated with abraded holder material. For example, fluorine was observable on GC after polishing in a Kel-F holder [134]. Fifth, polishing can change the surface O/C ratio. Rusling et al. [158] reported an increase from 0.12 to 0.17 with polishing, Cabaniss et al. reported a 0.18 O/C ratio on polished GC-20 [60], and Hu et al. reported 0.14 [83]. Rusling et al. also concluded that the majority of the oxygen on the surface is phenolic, and this species is increased during polishing. In addition, a depth profile of the polished surface revealed that polishing introduced oxygen to a depth of ca. 200 Å [158]. Hu, Karweik, and Kuwana used differential pulse voltammetry to establish the presence of orthoquinone groups on the polished surface [83]. Sixth, ultrasonic cleaning [160] or Soxhlet extraction [115,161] can substantially improve ET kinetics on polished surfaces, apparently by removing polishing debris and impurities. Polishing with diamond paste and no Al_2O_3 followed by Soxhlet extraction with toluene led to a large negative shift of E_p for AA, implying much faster ET kinetics [115]. Seventh, k^o for $Fe(CN)_6^{-3/-4}$ on polished GC decreased when the pH was increased from 4 to 7, while k^o for $Ru(NH_3)_6^{+2/+3}$ _increased_ over the same range [106]. Finally, and perhaps most importantly, the highest k^o values observed for polished GC surfaces are comparable to the highest values for Pt (Tables 6 and 9).

The polished GC surfaces producing the highest k_{obs}^o values for $Fe(CN)_6^{-3/-4}$ also show fast kinetics for ascorbic acid [117, 158] and dopamine [158]. Thus it is clear that proper polishing can produce large increases in ET rates for GC for several different benchmark systems. The question then arises of what is changing on the surface to bring about the activation. As noted in Sec. III, it could be surface cleaning, increased surface roughness, changes in the carbon microstructure, or surface oxidation. Perhaps polishing is producing active sites which are then deactivated slowly by solution impurities. Hu, Karweik, and Kuwana showed that both Pt and active GC electrodes degraded faster as the water purity decreased [83]. Since polishing may change so many variables, it is difficult to discern which is (are) most important.

Some insight into polishing effects are provided by ancillary experiments. The adsorption of 1,2,4-trihydroxybenzene was proportional to C_{obs}^o, implying that both are determined by either site density or microscopic area [162]. Raman spectra obtained on a fractured face of GC-30 before and after polishing are shown in Fig. 41 [163]. Note that the 1360 cm^{-1} band increased relative to the 1582 cm^{-1} band with polishing, implying a decrease in L_a during polishing. Since the Raman sampling depth is ca. 200 Å, the effects of polishing are fairly deep. Kazee and Kuwana presented evidence for a microparticle layer on polished GC, consisting of carbon particles and compacted polishing debris [164]. These experiments indicate that polishing changes the surface structure quite substantially from the bulk GC structure of Fig. 2, with the incorporation of oxygen, a decrease in L_a, and coating with an unknown level of impurities.

Several of the investigators of the polishing process propose that the "pristine" GC structure is important to fast ET [59, 157]. To pose the question differently, what is k^o for a pristine surface which has not been polished or otherwise pretreated?

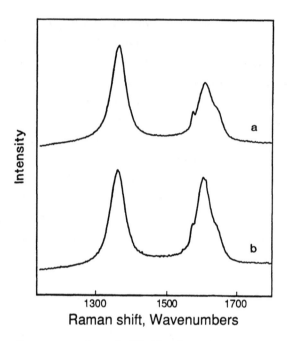

FIG. 41.   Raman spectra of GC-30 after fracturing (a) and polishing (b).   Note increase in 1360/1582 intensity ratio with polishing.   (From Ref. 163, reprinted with permission.)

When GC is fractured in solution, the surface is immediately exposed to electrolyte, without the involvement of polishing materials or any intentional oxidation [159,163].   The very high $k°$ of 0.5 cm/sec for $Fe(CN)_6^{-3/-4}$ on a fractured surface exceeds any GC results discussed so far, and also exceeds the best Pt values [159].   Although the fractured surface has not been characterized with respect to O/C ratio, wettability, etc., it is clear that the microstructural changes and other effects caused by polishing are not necessary to observe a high $k°$ for $Fe(CN)_6^{-3/-4}$.   AA and DA voltammetry on fractured GC also exhibit excellent ET kinetics [159,163].

To summarize, proper polishing of GC can result in a surface with ET activity comparable to Pt. The mechanism of activation by polishing is not clear since so many surface variables
change during polishing. However, the fast ET observed for
fractured GC implies that the underlying GC structure supports
fast ET, and polishing may activate by exposing the bulk GC
structure. The high variability in $k°$ on polished surfaces
probably indicates how well the polishing procedure prevents
impurity adsorption.

Although the main focus of work on polishing is the ET
rate, several polishing effects on background currents are available. The apparent double layer capacitance, measured with
AC techniques at 10 Hz or greater, yield values of 20 $\mu F/cm^2$
[83], 25 $\mu F/cm^2$ [165], 15 $\mu F/cm^2$ [166], and 24 $\mu F/cm^2$ [159].
When slower methods were used to determine C^o_{obs}, faradaic
reactions contribute to the current, yielding C^o_{obs} of 74 $\mu F/cm^2$
for GC polished on glass, measured by chronocoulometry [59].
As noted in Fig. 28, polished GC shows distinct peaks in a differential pulse voltammogram at ca. 175 mV and 700 mV vs.
Ag/AgCl at pH 2.0. The large observed "capacitance" on the
polished surface was attributed to surface redox groups, and
the differential pulse peaks were assigned to surface quinones.
Thus polished GC surfaces have a complex background consisting
of both double layer capacitance and faradaic contributions from
surface redox processes.

2. Heat Treatment Procedures

Although the effects of polishing on GC surfaces are far from
clear, heat treatment of the polished surface clarifies matters
substantially. Heating a polished GC electrode at 520–540°C in
ambient air produces an unusable surface with very high background current. Wightman et al. were the first to report that
heat treatment at reduced pressure (<1 torr) resulted in a major

increase in ET rates for AA, $Fe(CN)_6^{-3/-4}$, and catechols [161].
It was observed that the heat-treated electrodes remained active
for long periods of time in laboratory air, but that polishing in
alumina destroyed activity. Heat treatment at 725° in high vacuum ($<2 \times 10^{-6}$ torr) yielded a very active electrode for
$Fe(CN)_6^{-3/-4}$ for AA oxidation [59]. Based on the reported
ΔE_p for $Fe(CN)_6^{-3/-4}$ (1 M KCl) of 64.5 mV at 1 V/sec, $k°$
equals 0.14 cm/sec. E_p for AA at pH 2.0 was 0.260 V vs. SCE
[59]. Significant improvements in the reversibility of DOPAC
and DHBA were observed as well [161]. Thus vacuum heat
treatment (VHT) resulted in ET kinetics as fast as that observed
for the best polishing procedures, and polishing was deleterious
to VHT-induced activation.

While the effect of VHT on ET kinetics is apparently similar
to ultraclean polishing for AA, catechols, and $Fe(CN)_6^{-3/-4}$,
the effect on other surface properties is dramatically different.
VHT causes the surface O/C ratio to decrease from 0.22
(polished) to 0.06 after VHT [59]. The peaks in the differential pulse voltammetry peaks attributed to surface quinones are
absent for the VHT surface (Fig. 28). A chronocoulometric
measurement of capacitance of a VHT surface yielded 10 $\mu F/cm^2$
[59]. These results indicate that VHT surfaces are low in oxygen content and background current, yet exhibit fast ET kinetics. Although there is a possibility of oxidation of the GC
after removal from the vacuum, it is clear that neither electroactive surface oxides nor high surface roughness are prerequisites for fast ET for AA and $Fe(CN)_6^{-3/-4}$.

3. Laser Activation of GC

Laser activation (LA) is a relatively recent addition to the collection of GC pretreatments and is in some ways similar to VHT.
As noted earlier for HOPG, a short, intense laser pulse impinges
on a GC surface either in air, an inert atmosphere, or directly
in the electrochemical cell [126–128]. The laser is usually a

Nd:YAG operating at 1064 nm, with a 9-nsec pulse duration. Nitrogen [127,167] and iodine [168] lasers have also been used successfully. Because GC absorbs more laser light than HOPG, the power density required for activation is lower, about 20–30 mW/cm^2 [126,159]. The laser pulse results in very rapid heating of the surface to 1000°C or more [131,169]. The activation of electron transfer is weakly dependent on the laser wavelength, implying that the activation mechanism is thermal rather than photochemical [169]. The beneficial effects of LA for the voltammetry of AA and DA are demonstrated by Fig. 42.

Although both VHT and laser activation may be thermal processes, there are some important distinctions. Laser activation is very fast, requiring 9 nsec, and repeatable, at the repetition rate of the laser (10 Hz for a typical Nd:YAG laser). LA can be performed in situ, with no solvent removal or cell disassembly. LA can be synchronized with an electrochemical measurement, e.g., at some particular time relative to a potential pulse. In the case of differential pulse polarography, LA was used to pretreat the electrode before each potential pulse, thus preventing electrode fouling and maintaining ET activity [127]. As we saw in Sec. IV.A.2, laser activation decreases L_a for HOPG, indicating that LA can fracture a graphite lattice.

As shown in Fig. 43, LA can result in a large improvement in ET kinetics for DA over a conventionally polished surface [159], but the polished, laser-activated surface is only slightly better than a fractured surface. The high activity of an LA surface lasts for several hours, with the lifetime shorter for lower purity electrolyte [126], very similar to the behavior observed for properly polished electrodes [83]. The lifetime is shorter in an LC flow cell [170], consistent with the conclusion that loss of ET activity is caused by adsorption of impurities from solution. The effectiveness of LA for several quinones is demonstrated by the decrease of ΔE_p shown in Table 10. LA

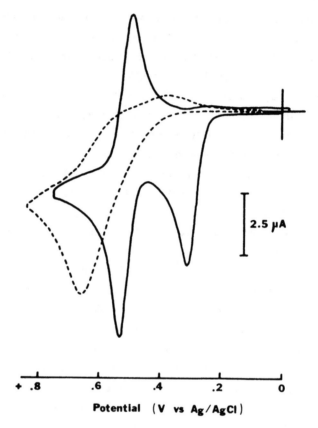

FIG. 42.   Voltammetry of a mixture of DA and AA in 0.1 M
H$_2$SO$_4$.   Dashed line is polished surface, solid is after in situ
laser activation.   (From Ref. 126, reprinted with permission.)

is compared to several pretreatment protocols for the case of
AA oxidation in Fig. 44.   Note that laser activation in solution
leads to consistent E$_p$ values close to the thermodynamic limit.
In addition, the E$_p$ values from LA are comparable to those from
VHT or the best polishing procedures.

For Fe(CN)$_6^{-3/-4}$, the k$_{obs}^o$ on LA GC is approximately at
the upper limit of the techniques used to measure it.   Fast

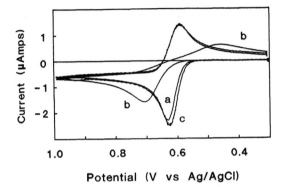

FIG. 43.   Voltammetry of DA on GC-30 in 0.1 M H_2SO_4 (ν = 1 V/sec).   Curve a is fractured surface, b is after polishing, and c is after laser activation at 25 MW/cm^2.   (From Ref. 159, reprinted with permission.)

TABLE 10

Laser Activation of Electron Transfer

Redox system	ΔE_p, polished[a,b]	ΔE_p, laser activated[c]
$Fe(CN)_6^{-3/-4}$	0.130 V	0.06 V
Dopamine	0.125	0.032
DOPAC	0.160	0.027
DHBA	0.235	0.030
Catechol	0.158	0.025
Hydroquinone	0.113	0.048

[a]1 M KCl, 0.1 M phosphate, pH 7.0, ν = 0.1 V/sec.
[b]Al_2O_3 on polishing cloth.
[c]Three 20 MW/cm^2 pulses, in situ.
Source: From Ref. 126.

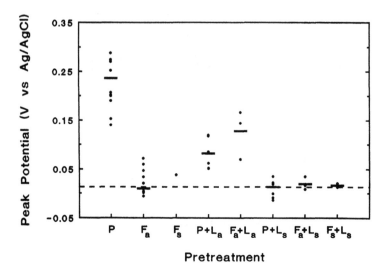

FIG. 44. Observed anodic peak potential for ascorbic acid in
pH 7.0 phosphate buffer, following various pretreatments.
Points indicate individual experiments, bars are averages,
dashed line is E_p on a mercury electrode. Abbreviations: P,
polished; F_a, fractured in air; F_s, fractured in solution; L_a,
laser activated in air; L_s, laser activated in solution. (From
Ref. 159, reprinted with permission.)

scan rates are required to measure high $k°$ values, and uncom-
pensated resistance becomes a serious problem. When GC electrodes are made as small as practical (ca. 0.5×0.5 mm) scan
rates up to about 100 V/sec can be used without significant
ohmic error. Figure 45 shows cyclic voltammograms for
$Fe(CN)_6^{-3/-4}$ on laser-activated GC. Curves simulated for $k° =$
0.005, 02, 0.5, and 0.8 cm/sec are shown; above 0.8 cm/sec
the voltammograms are indistinguishable. The experimental
curve overlaps the curve simulated for a $k° = 0.8$ cm/sec [159].
For eight repetitions of polishing and LA, $k°$ was always above
0.5 cm/sec. Although double layer effects could easily account
for a factor of 2, it is clear that $k°$ for $Fe(CN)_6^{-3/-4}$ on a LA

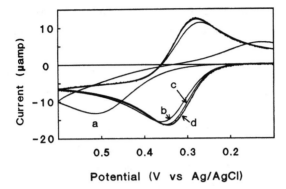

FIG. 45. Simulated and observed voltammograms for 1 mM
$K_4Fe(CN)_6$ in 1 M KCl on polished, laser activated GC-30, following 18 25 mW cm^{-2} laser pulses. Smooth curves were simulated using standard explicit finite difference techniques for
$\alpha = 0.5$, $\nu = 100$ V/sec, and k^o as follows: curve a) 0.005;
b) 0.2; c) 0.5; d) 0.8 cm/sec. Noisy curve overlapping curve
d is the background subtracted experimental voltammogram.
Vertical axis of experimental curve was adjusted to match curve
d. (From Ref. 159, reprinted with permission.)

surface is at least as high as the 0.24 observed for the best Pt
surfaces. In addition, there is no evidence for formation of a
passivating film from $Fe(CN)_6^{-3/-4}$ on GC. Contrary to the case
for polished GC [108], the k^o for $Fe(CN)_6^{-3/-4}$ on the LA surface
is not pH dependent [128].

SEMs of the GC surface show minor changes upon LA. As
was observed with VHT [164], LA reveals polishing scratches,
perhaps by removing a superficial layer of polishing debris.
The surface O/C ratio decreased upon LA in UHV, until the
surface oxygen was undetectable (<1%) by ESCA [126]. LA
reduced the O/C ratio by an average of 38% when LA was carried out in aqueous solution [126]. Like VHT, LA leads to a
surface low in oxygen but electrochemically active. The AC
capacitance (100 Hz) for a LA surface is 29 $\mu F/cm^2$ [159],

slightly higher than that observed for polished surfaces. The
apparent capacitance measured by the slower chronocoulometric
technique increases with LA, with the magnitude of the increase
depending on conditions. LA at high power density in solution
can lead to C_{obs}^o of 200 $\mu F/cm^2$, while LA in air at 10–80 MW/
cm^2 yields low observed capacitance, ca. 10–15 $\mu F/cm^2$ [128].
Capacitance and k^o enhancement were decoupled for LA, and
fast k^o could be observed with only slight increases in C_{obs}^o.
Finally, adsorption of PQ is not altered significantly by LA.
Γ_{obs} for PQ increased by less than a factor of 2.0 upon LA of
a polished surface [159]. Adsorption of 2,6-diamino-8-purinol
occurred on LA surfaces, but did not change significantly with
power density [167]. In the same study, it was concluded that
LA did not significantly alter microscopic surface area, and that
observed capacitance did not correlate with ET activation. It
has also been noted that weak DA adsorption occurs on surfaces
laser activated in solution, but not those activated in air [128].
ET activation occurred whether or not DA adsorption increased.

In summary, laser activation of GC results in the highest
k^o yet observed for $Fe(CN)_6^{-3/-4}$, and fast ET for AA and DA.
LA in solution leads to moderate increases in capacitance and
adsorption, which can be severe at high power densities. However, LA in air does not lead to such increases, and changes
in adsorption are not necessary to observe ET activation. In
addition, ET activation is accompanied by small changes in observed capacitance. For example, a capacitance increase of
about a factor of 2 accompanies an increase of k^o by a factor
of 100. Despite major differences in technique, laser activation
and VHT produce functionally similar surfaces, with fast ET
kinetics but relatively low background currents. Both approaches lead to the conclusion that microscopic roughness is
not significantly increased during activation, and cannot account

for the increased $k°$. In both cases activation is at least partially attributable to improved surface cleanliness.

4. Electrochemical Pretreatment (ECP) of Glassy Carbon
ECP has been studied in some detail for GC, carbon fibers, and
carbon composite electrodes, because it is quite effective toward
improving electrode activity. One of the earliest examples of
ECP [171,172] involved pretreatment of carbon fiber microelectrodes for in vivo analysis in brain tissue. ECP improved ET
kinetics sufficiently to separate the previously overlapping AA
and DA voltammetric peaks, and thereby permit detection of DA
in vivo. In addition, the analytical sensitivity for DA was at
least an order of magnitude higher than that for AA. The mechanism of ET activation and the enhanced selectivity for DA stimulated research in several laboratories.

The variables in ECP of GC electrodes include electrolyte
composition, applied potential waveform, and magnitude of applied
potential. As with polishing procedures, there is no consensus
on the most effective ECP procedure, and most laboratories adapt
the procedure to suit the intended use of the electrode. The
effects of ECP on carbon fiber electrodes are often different
from those on GC, due to the much higher current density.
This section will deal only with "large" GC electrodes, and fibers will be covered later.

The most thoroughly studied ECP was devised by Engstrom
[56,134] and involves a dc potential near 2.0 V vs. SCE, usually
followed by a reduction step. Several laboratories have reported
on the effects of this pretreatment or minor variants on ET kinetics and surface structure. Engstrom and Strasser [134] oxidized a polished GC electrode for 5 minutes at 1.75 V vs. SCE
in 0.1 M KNO_3, followed by reduction at -1.0 V for 1 min. The
ECP treatment had a major effect on the $E_{1/2}$ for rotating disk
voltammetry of several systems, some of which are listed in
Table 11. The $E_{1/2}$ for $Fe(CN)_6^{-3/-4}$ shifted by only a few

TABLE 11

Effect of ECP on Electron Transfer Kinetics for GC Electrodes

Procedure	Redox system	$E_{1/2}$[a], initial	$E_{1/2}$[a], after ECP	Ref.
1.75 V dc, −1 V dc 0.1 M KNO$_3$	hydroquinone	0.332 V	0.187 V	134
	catechol	0.391	0.323	
	AA	0.165	0.060	
	Fe(CN)$_6^{-3/-4}$	0.172	0.166	
	Fe$^{+3/+2}$	0.016	0.291	
	O$_2$	−0.68	−0.31	
		ΔE_ρ[b], initial	ΔE_ρ[b] after ECP	
1.8 V dc, 28 min, −0.2 V dc, 0.1 M H$_2$SO$_4$	catechol	0.445 V	0.095 V	60
	(NH$_3$)$_5$Ru^{IV}O	0.610	0.095	60
Potential cycling, 0.24−2.04 V, 0.1 M H$_2$SO$_4$	catechol	0.250	0.040	55

[a]From RDE voltammogram.
[b]From cyclic voltammetry.

mV upon ECP. Cabannis et al. found that k^o_{obs} for both polished and ECP electrodes exhibited a kinetic isotope effect, with
the kinetics in H_2O being faster than that in D_2O for several
redox reactions which involve protons, e.g., H_2O_2 oxidation
[60]. They propose that H^+ is involved in the transition state
for electron transfer, and that ET is accompanied by H^+ transfer
to or from the GC surface. For example, the reaction:

$$(bpy)_2(Py) \; Ru^{IV} \; (OH)^{+2} \rightarrow (bpy)_2(py) \; Ru^{III}(O)^{+2} + e^- + H^+$$

(26)

requires proton transfer for completion. If both the proton and
electron are involved in the rate-limiting step, there should be
an H/D isotope effect. Such an effect was observed, and the
authors propose that H^+ is provided by oxygen functional groups
on the activated GC surface. Nagaoka et al. concluded that ECP
activation of O_2 reduction did not involve redox mediation by
surface quinones [89].

The ET activity improvement from ECP is robust and well
established, and led to several investigations of the oxide film
observed on the electrode after ECP. Visually, the electrode
changes color (pink, green, then gold) [56,134], an effect attributed to interference effects in a semitransparent film. Beilby
noted that color changes did not occur when ECP was conducted
in 1 M NaOH [173]. XPS analysis of the dried oxide film revealed an increase in the O/C ratio [60,134], and elemental
analysis indicates that about 25% of the dried film is oxygen
[55]. High resolution ESCA revealed that much of the oxygen
is in the form of phenolic or alcoholic functional groups [60,174],
and a smaller amount occurs as carboxylate [174]. Carbonyl
groups were detected in the oxidized film but disappeared upon
reduction [174]. Ellipsometry of the GC surface during potential cycling in 0.1 M H_2SO_4 (0.0–2.04 V vs. SCE) showed
growth of a transparent film at a constant rate of about 45 Å/

cycle, and that film growth continued until its thickness was at least 1 μm [55]. The film is porous, hydrated, and nonconductive, and structurally similar to graphitic oxide [55,173]. Electron transfer rates to quinones improved up to a film thickness of 250 Å, with thicker films yielding no additional improvement [55]. Electrodes pretreated either with constant or cycled potentials exhibited surface redox waves usually attributed to surface bound quinones. In addition, GC electrodes have significantly higher background current following ECP. This background is much higher after acidic ECP than basic ECP. ECP may be used to renew GC electrodes in situ [175] and in LC detectors [176], resulting in useful electroanalytical procedures.

The adsorption properties of the oxide film formed by ECP are unusual, and have been the subject of several investigations. The GC surface adsorbs ions following ECP [88] and can undergo ion exchange of cations. Given the XPS evidence for carboxylate groups on anodized GC, the ion exchange may involve protons from carboxylic acids or even phenolic groups in the EGO film. Catechol [87], dopamine [173,177], 1,4-naphthoquinone [128], and nitrobenzene [178] all absorb strongly at electrodes electrochemically pretreated in acid or neutral solutions. Ascorbic acid exhibits adsorption on ECP glassy carbon in 0.1 M H_2SO_4, but not at pH 7.0 [128]. Anjo et al. found that DA adsorption increased with ECP time and potential, as expected, and that an activation potential of 1.2 V was required to reduce ΔE_p fully for DA [177]. However, if ECP was carried out at high pH (e.g., 1.0 M NaOH) or if an acid-activated GC surface was washed with 1.0 M NaOH, DA adsorption was greatly decreased [173,177]. For example, ECP of GC at pH 1.0 yielded DA adsorption of 2.5×10^{-9} moles/cm^2, while activation in 1.0 M NaOH yielded 3.1×10^{-10} moles/cm^2 [177]. It is important to note that activation in base yielded excellent ET activation (i.e., small ΔE_p) for DA, yet the resulting surface exhibited

relatively weak DA adsorption. Finally, Poon, McCreery, and
Engstrom concluded that ECP yields surfaces with fast ET kinetics, relatively high background current, and high adsorption,
while laser-activated surfaces exhibit fast ET kinetics for DA,
$Fe(CN)_6^{-3/-4}$, and AA, but relatively low background current
and adsorption [128]. Furthermore, the high background and
adsorption could be greatly reduced by laser irradiation, while
retaining fast ET kinetics. Qualitative comparisons of several
pretreatment procedures are listed in Table 12.

Although the several studies cited here use differing ECP
conditions and in many cases different test redox systems, there
appears to be agreement on several points. First, ECP results
in activation of a variety of redox reactions, particularly catechols and ascorbic acid, and to a lesser extent $Fe(CN)_6^{-3/-4}$.
Second, ECP in acid leads to a large increase in background
current, with a surface redox couple similar to a quinone usually
observed in background voltammograms. Third, ECP in acid
yields a film which adsorbs DA or catechol and can ion exchange
with the solution. Fourth, XPS and elemental analysis indicate
an increase in O/C ratio with ECP, with an increase in alcoholic/
phenolic groups most prominent. Fifth, the oxide film generated
with ECP increases in thickness with electrolysis time, and is
usually several hundred Å up to 1 μm thick. Sixth, ECP removes polishing impurities from the GC surface.

As noted in Sec. IV.A.3, it is difficult to postulate a structure for EGO, but some inferences may be drawn. It is clearly
porous and permeable to ions and water. Adsorption of DA
(a cation) and ascorbic acid (neutral) are easily observed, but
adsorption of ascorbate (an anion) is weak or absent. DA adsorption and voltammetric background are greatly reduced by
treatment of an ECP electrode in base. The observations are
consistent with the model shown in Fig. 46. A conventionally
polished GC surface has a layer of polishing debris, impurities,

TABLE 12

Comparison of Pretreatment Procedures for GC

	ET activation	Adsorption	Background current	Ref.
ECP, low pH	yes	high	high	56,87,128, 173,177
ECP, high pH	yes	low	moderate	173,177
VHT	yes	low	very low	59
Laser activation in air	yes	low	low	128,167
Conventional polishing	no	low	moderate	128
Ultraclean polishing	yes	moderate	moderate	83

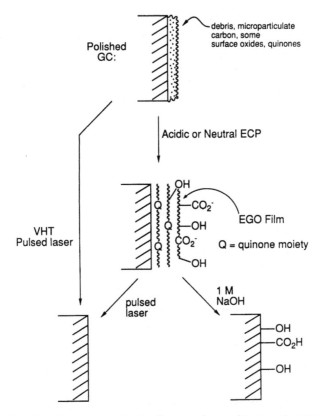

FIG. 46. Working hypothesis for surface effects of ECP, VHT, and laser activation on polished GC. All three procedures remove polishing debris, but ECP produces a thick oxide film. Treatment with base or ECP in base removes the oxide film, but may leave many surface oxides. VHT and laser-activated surface have low background current and apparently low surface oxide level.

and perhaps microparticulate carbon which inhibit electron transfer. Most polished surfaces exhibit surface redox waves and have an O/C ratio of roughly 0.15. When the polished surface is anodized, the nonconducting but porous graphitic oxide layer is formed and some or all of the polishing debris is removed. CO_2 or O_2 may be evolved, further aiding the removal of impurities. The oxide layer contains quinone redox centers, as well as hydroxyl groups and to a lesser extent carboxylates. It is not the ideal $C_8(OH)_4$ structure shown in Fig. 16, but probably contains the hydrated nonconductive fused cyclohexyl backbone. The EGO modified GC electrode has high background current caused by redox processes in the film, or perhaps by a thin conductive region with a high microscopic surface area. The film adsorbs catechols, particularly cations, due to anionic sites in the film. The enhanced sensitivity for DA over AA at pH 7 [154,155,172] is probably due to differential adsorption or permeation of cations vs. anions. If the oxide film is removed by exposure to base [177] or by a pulsed laser [128], the adsorption and higher background are greatly reduced, and surface quinones are no longer evident. Fast electron transfer kinetics are retained after NaOH treatment, indicating that high adsorption and background current are not prerequisites of high ET activity. It is clear from VHT and laser activation that surfaces with immeasurably low surface O/C ratios have fast ET to $Fe(CN)_6^{-3/-4}$, AA, and DA. Thus measurable surface oxygen is not necessary for fast ET to these systems, but it does promote adsorption and background current if present. One concludes from the available results that ECP of GC cleans the surface of many impurities, and that oxidation per se is not necessary for ET activation of $Fe(CN)_6^{-3/-4}$, AA, and DA. Several arguments against the involvement of oxides in ET to these systems have been made [135].

The noninvolvement of surface oxygen in ET is not general to all systems. There are well-established cases of redox mediation by surface quinones [179,180]. An example is shown in Fig. 47, in which NADH is catalytically oxidized by a surface bound quinone. The proton-assisted electron transfer mechanism based on the H/D isotope effect requires surface oxides, particularly phenols [60]. These reactions can be classified as inner sphere or electrocatalytic, since the carbon surface is involved chemically, as more than a source or sink of electrons. The ECP results permit the conclusion that an oxide film may be necessary for certain electrocatalytic reactions, but that many redox systems such as $Fe(CN)_6^{-3/-4}$, AA, and DA, undergo fast electron transfer with carbon in the absence of observable oxide films.

E. Carbon Fiber Electrodes

Carbon fibers have been studied in some detail, not only because of applications to in vivo analysis, but also because they can be made into disk or cylindrical microelectrodes which exhibit nonplanar diffusion. As noted in Sec. II, carbon fibers

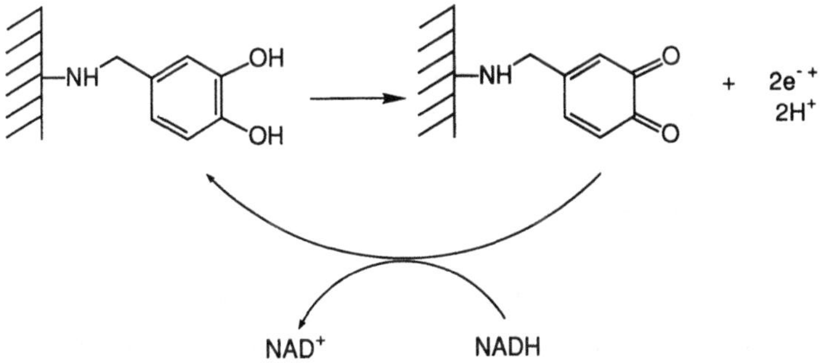

FIG. 47. Example of redox mediation by a surface-bound orthoquinone. Overall reaction is NADH oxidation mediated by electrochemical oxidation of catechol (see Ref. 180).

vary greatly in microstructure, and are to some degree anisotropic, with the a-axis of the graphitic lattice preferentially
oriented along the fiber axis. In addition, the onion and to some
extent random models of Fig. 12 have higher concentrations of
basal plane on the side of the fiber than on the ends. Given
the large differences in electrochemical behavior for edge vs.
basal plane, we would expect different electrochemical responses
for fiber ends vs. fiber sides. Furthermore, the wide variety
of synthesis and heat treatment procedures would be expected
to lead to a wide range of microstructure and electrochemical
behavior.

The small diameter of carbon fibers (typically 5–25 μm)
leads to some unusual behavior, and applications of particular
note are intracellular and in vivo analysis of neurotransmitters
(e.g., dopamine) [172,181] and microanalysis of chromatographic
eluents [182,183]. Although these experiments are important
and ingenious, our main interest here is in electron transfer
and background processes, and the applications of carbon fiber
electrodes will not be discussed in any detail. Of the various
electrode preparation approaches, we will cover only electrochemical pretreatment and the properties of unmodified fibers.

Many commercially available carbon fibers have been coated
or oxidized to improve adhesion in composite materials, and care
should be taken to avoid or remove such coatings. Adams et
al. [184] compared the capacitance of the end of the carbon
fiber (as a disk electrode) with that of the exposed cylinder.
Based on voltammetric background current and geometric area,
the untreated fiber had a C_{obs}^{o} of 23 $\mu F/cm^2$ on its end and
6.8 $\mu F/cm^2$ on the cylindrical surface. This difference is
caused either by increased roughness on the fiber end or by a
higher fraction of basal plane on the cylindrical surface. Since
the pure basal plane capacitance is 2–3 $\mu F/cm^2$, the low cylindrical C_{obs}^{o} could be a consequence of fiber microstructure.

With a few notable exceptions, voltammetry at untreated fibers usually shows slow ET and often very poorly defined waves. In an early study by Ponchon et al., untreated fibers were used for differential and normal pulse voltammetry of catecholamines and other neurotransmitters [171]. Although they reported micromolar detection limits for DA in aqueous solutions, the electron transfer and adsorption properties of the fiber were not discussed. An elliptical disk electrode formed by the exposed end of a glass-encased carbon fiber has been successfully used for in vivo analysis and characterized in detail [185]. After the carbon fiber was sealed in a glass capillary, the capillary was cut and beveled at an angle with 1 μm diamond paste. The result was an ellipse of exposed carbon with a minor axis of 10 μm, and no further physical or electrochemical pretreatment was performed. Fast scan voltammograms (200 V/sec) of several neurotransmitters and related compounds are shown in Fig. 48. Although all redox reactions were quasireversible at the high scan rate, it was clear that the electrode was much more sensitive to cationic analytes than neutral or anionic. For example, 200 μM ascorbic acid (an anion at pH 7.4) yielded a wave less than one-third as large as 10 μM dopamine (a cation), implying a relative sensitivity of more than 60:1 for DA over AA. The effect was even greater when the electrode was coated with Nafion as a cation-permeable membrane. Untreated electrodes exhibited a surface-bound background wave similar to that for surface quinones [185]. Analysis of the voltammograms revealed that the cationic catechols were adsorbed on the carbon surface and the majority of the voltammetric response at high scan rates was due to adsorbed species. Thus the enchanced response of cations over neutrals or anions was due to adsorption to a carbon surface which had only been polished with diamond paste. The adsorption was much less obvious at lower scan rates [186].

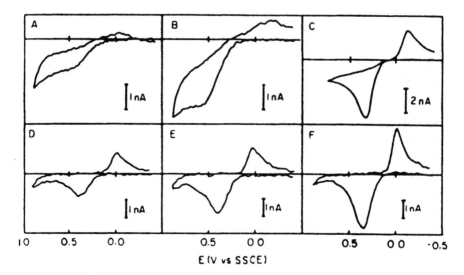

FIG. 48. Fast scan (200 V/sec) voltammograms of AA and several catechols on polished carbon fiber disk electrodes. All in pH 7.4 buffer, and background voltammograms were subtracted. Redox systems were: (A) 200 μM AA; (B) 200 μM DOPAC; (C) 100 μM 4-MC; (D) 10 μM NE; (E) 10 μM DHBA; (F) 10 μM DA. (From Ref. 185, reprinted with permission.)

On a related point, DA adsorption is not apparent on a fractured GC surface, based on semiintegration of a 1 V/sec voltammogram [187].

Electrochemical pretreatments of carbon fibers fall into two distinct groups, referred to here as "mild" and "strong." Mild treatment involves potential sweeps or steps from 0.0 to 1.3–2.0 V vs. SCE, often in phosphate-buffered saline, PBS (i.e., 0.15 M NaCl + phosphate buffer, pH 7.4). This "mild" treatment produces significant enhancement of ET kinetics, similar to the ECP of GC. A "strong" treatment involves maximum potentials of 2.5–3.0 V vs. SCE, and results in fibers with distinct properties.

"Mild" ECP improved fiber performance, particularly for catechols [184,188,189]. Michael and Justice [188] found that DA adsorbed on the exposed end of a fiber after mild ECP, but 4-MC did not. DA adsorption was not apparent in voltammograms at slow scan rates (0.1 V/sec) but became pronounced at 10 V/sec. Similarly, Sujaritvanichpong et al. found that DA oxidation was diffusion controlled for long chronoamperometric experiments (>0.1 sec), but adsorption was apparent at short times [190]. Particularly convincing evidence for DA adsorption is the persistence of a DA voltammetric wave after washing with blank electrolyte [188].

Thus "mild" ECP improves ET performance of untreated carbon fibers [184,188,190], and significantly increases their adsorption of cationic catechols. Since carbon fibers are difficult to clean mechanically (e.g., by polishing), the cleaning process observed for GC may be of major importance to the mechanism of "mild" ECP. The adsorption of DA exhibited by the untreated fiber appears to be quantitatively greater after mild ECP, but qualitatively similar.

One of the seminal papers on carbon fiber electrochemistry involved "strong" pretreatment with high potentials in the range from 2–3 V vs. SCE [172]. Gonon et al. [191] found that after pretreatment with a 70 Hz, 0–2.8 V triangular wave, the differential pulse voltammetry of AA and DA revealed peaks at the reversible potentials, implying a dramatic increase in ET rates. Furthermore, the sensitivities for analytes relevant to neurochemical analysis varied greatly. Relative sensitivities for DA, DOPAC, and AA were about 1000:10:1 at pH 7.4. In addition, DA was detectable at about 50×10^{-9} M, indicating outstanding electroanalytical performance. The "strong" ECP resulted in nearly ideal electrodes for in vivo monitoring of DA, since the response to major interferents was greatly suppressed.

The origin of this favorable development was revealed by investigations from several laboratories, with several observations providing clues. First, the capacitance of the carbon increases greatly with ECP, to about 120 $\mu F/cm^2$ [184]. Second, the voltammetric sensitivity for $Fe(CN)_6^{-3/-4}$ decreased somewhat with ECP, but that for DA increased significantly [184]. Voltammograms for $Fe(CN)_6^{-3/-4}$ exhibited currents which were smaller than expected for the exposed area of the fiber. Third, as shown in Fig. 49, ECP increased sensitivity for DA (a cation), had little effect on 4-MC (a neutral), and decreased sensitivity for AA and DOPAC (anions at pH 7.4) [192]. Fourth, the oxidation current for DA responded slowly to changes in DA concentration, and the response decayed with repeated scans [184, 190]. Fifth, DA was strongly adsorbed on the ECP fiber, at a surface concentration consistent with multilayer adsorption [184, 190,192]. Sixth, if the "strong" ECP potential was too high (>3.0 V), all responses for AA, DA, etc., decreased dramatically. Finally, the response to inorganic and organic cations is greatly enhanced by strong ECP, but that of anions is suppressed, as shown in Fig. 50 [192].

Based on comparisons of theory and experiments for a variety of redox systems, Kovach, Deakin, and Wightman proposed a mechanism for the effects of "strong" ECP [192]. The current density during ECP reaches 2–3 A/cm^2, since the electrodes are small and the potential high. This represents a much more vigorous oxidation than the "mild" treatment and leads to the distinctive surface properties. Kovach et al. conclude that an insulating EGO layer partially covers the fiber. DA and other cations can ion exchange with the EGO film, thereby preconcentrating on the electrode. The active electrode areas are microcracks in the EGO film which are ca. 0.7 μm wide on a 5-μm diameter electrode. These cracks exhibit fast ET to anions or cations. $Fe(CN)_6^{-3/-4}$ does not adsorb in the EGO film, but

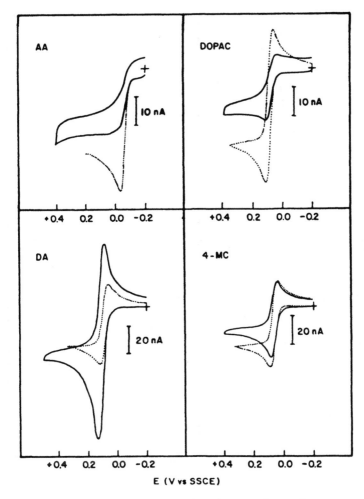

FIG. 49.   Voltammetry at the cylindrical surface of a carbon
fiber following "strong" pretreatment, all at pH 7.4.   Solid
lines are experimental data, dashed are simulated assuming fast
k^{o}.   Concentrations were 0.2 mM for DA, DOPAC, and 4-MC,
1.0 mM for AA.   (From Ref. 192, reprinted with permission.)

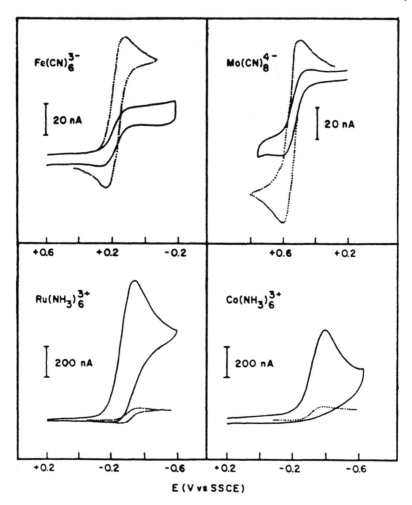

FIG. 50. Voltammetry of several metal complexes on a carbon fiber following "strong" ECP. Solid line is experimental, dotted is simulated. Concentrations were 1.0 mM in all cases, solution pH was 7.4. (From Ref. 192, reprinted with permission.)

it does exhibit voltammograms predicted for a 0.7-μm wide hemi-cylindrical electrode. The active electrode surface is only about 7% of the total electrode area, with the remainder consisting of insulating EGO. DA and other cations have large responses because of preconcentration in the film, while AA and anions are suppressed. Since DA adsorption occurs via ion exchange into a multilayer film, much more than a monolayer of DA may adsorb [184,190,192]. The slow response to changes in DA concentration is caused by the time required to permeate the EGO film and preconcentrate the DA. As pointed out by Feng et al., there is a trade-off betwen response time and sensitivity, since thicker EGO films permit lower DA concentrations to be observed [184].

Relatively few experiments have been performed on the effect of carbon fiber microstructure on electrochemical performance. Feng et al. [184] noted that the cylindrical surface of a fiber exhibited poor electrochemical performance compared to the fiber end, but both improved greatly upon ECP. This observation is consistent with the greater basal plane character usually present on the side of the fiber (see "random" and "onion" cases, Fig. 12). Wightman et al. [115] noted that low modulus fibers exhibited faster ET kinetics than high modulus fibers, because the latter had a smaller fraction of exposed edge plane. While these observations do not represent a systematic study, they do emphasize the importance of a-axis orientation and edge plane exposure to electrode characteristics.

F. Carbon Film Electrodes

As discussed in Sec. II, carbon films may be fabricated by vapor deposition, sputtering, or pyrolysis of hydrocarbons, and the degree of microstructural order depends strongly on temperature. The most common films are prepared for electrochemistry by pyrolysis, usually in a temperature range of 900–1200°C.

Formation of pyrolytic carbon films has a major advantage of
being adaptable to a variety of electrode shapes and sizes. A
pyrolytic carbon film can coat a complex shape or a very small
structure such as a quartz micropipette. For example, a pyrolytic carbon ring electrode has been made with overall diameter in
the 1–5 μm range [193,194]. In some cases lower deposition
temperatures were used, either by catalyzing the deposition with
a nickel surface at 400–550°C [195] or by using a less stable
precursor such as perylene tetracarboxylic acid, treated at 850°C
[196]. A low formation temperature for a carbon film may have
significant practical advantages, particularly for less refractory
substrates such as glass. In addition to these pyrolytic materials, a commercial low temperature isotropic carbon (LTIC) has
been evaluated as an LC detector [197,198]. Its structure was
not discussed in any detail.

The carbon microstructure of carbon films used for electrochemistry is rarely characterized, and the films could range in
structure and properties from semiconducting amorphous carbon
to pyrolytic graphite. However, in most cases the films reported
in the literature behave much like disordered PG. Several examples that bear on structure/reactivity concerns of carbon
electrodes will be noted here. Most carbon film electrodes exhibit moderate electron transfer characteristics without surface
preparation. Before any ECP, Lundstrom [199] reported a ΔE_p
of 80 mV for $Fe(CN)_6^{-3/-4}$ (2 M KCl) at 10 mV/sec, on a film
electrode prepared at 1100°C. The k° estimated from this ΔE_p
is 0.004 cm/sec, which is comparable to the value of 0.006 reported by Beilby and Carlsson [189].

As with carbon fiber electrodes, a variety of electrochemical
pretreatments have been applied to improve ET kinetics. As
shown in Table 13, k_{obs}^o for $Fe(CN)_6^{-3/-4}$ increased significantly
after ECP, with rates approaching those of properly prepared
GC. The effect was more dramatic for dopamine, for which the

TABLE 13

k° for $Fe(CN)_6^{-3/-4}$ on Pyrolytic Carbon Film Electrodes

Conditions	Film preparation temperature	k° before ECP	k° after ECP	Ref.
0.1 M KCl	natural gas/O_2		1.6×10^{-3} cm/sec	200
1 M KCl	1075°C	0.006 cm/sec	0.05	189
1 M KCl	1025°C	0.005	0.06	189
1 M KCl	975°C	0.02	0.1	189
2 M KCl	1100°C	0.004	0.002	199
0.1 M NaNO₃ 0.1 M Phosphate	850°		0.001	196
pH 7				
2 M KCl	LTIC		0.012	197

ΔE_p decreased from about 0.35 to 0.05 volts after anodization of
the electrode for 5 min at 1.3 V vs. SCE [199]. Interestingly,
nearly the same improvement of DA ET kinetics occurred when
the pyrolytic film was washed with dichloromethane. Lundstrom
concluded that surface cleaning was largely responsible for the
decrease in ΔE_p, but that a small additional decrease occurred
when the electrode was anodized [199].

The capacitance of carbon films varies with preparation; examples are 20 $\mu F/cm^2$ for an 1100° film [199] and 7–10 $\mu F/cm^2$
for 975–1075° films [189], all based on voltammetric background
current. C_{obs}^o always increased with ECP, with values of 29–63
$\mu F/cm^2$ observed after treatment. Since anodization will increase
the density of surface oxides, C_{obs}^o is likely to include a surface
faradaic component.

When the carbon film is deposited onto a small structure
such as the inside of a quartz capillary [193], some unusual
effects were observed following ECP. For a 5–10 μm diameter
ring made from methane heated by a methane/O_2 flame, cationic
catechol derivatives such as DA exhibited greatly increased re-
versitility, neutral catechols showed little change and anionic
catechols (e.g., DOPAC) showed decreased ET rate. For the
same treatment on carbon fibers, both cationic and anionic catechols exhibited improved kinetics. The authors ruled out a
double layer effect and concluded that an anionic surface site
on the film must be specifically interacting with cationic catechols,
and that there is a significant difference in the effect between
carbon films and carbon fibers.

Although there are differences in procedures and conditions,
the observations on carbon films are consistent with previous
discussions on GC and carbon fibers. As deposited films have
fairly low C_{obs}^o and poor kinetics, probably because of surface
impurities. ECP increases C_{obs}^o significantly, apparently by
forming surface oxides, and also increases k^o, probably by

removing impurities. As shown with the improvement of $k°$ following a CH_2Cl_2 wash, surface oxidation is not a prerequisite for
high $k°$ for DA. Surface oxidation can cause specific chemical
interactions, particularly between a cationic catechol and a surface bound anionic group. Based on available observations,
one would predict that vacuum heat treatment or laser activation
will yield fast ET kinetics for $Fe(CN)_6^{-3/-4}$ on carbon films.

V. CONCLUSIONS

In order to draw generalizations about the relationship between
carbon electrode structure and ET reactivity, one must overcome
the wide variation of carbon microstructures, redox systems,
and pretreatments. Nevertheless, there are some useful conclusions available from the work cited here which permit insights
into ET mechanisms and have predictive value when considering
a new carbon material or pretreatment. In this section, the
main points will be summarized and a working hypothesis for the
relationship between surface structure and ET reactivity will be
formulated. Comments are subdivided according to ET reactivity,
capacitance, and adsorption behavior.

A. Electron Transfer Reactivity

The many heterogeneous redox reactions studied with carbon
electrodes can be subdivided into two groups which differ in
their sensitivity to surface oxides. Those with a $k°_{obs}$ which
is uncorrelated with surface oxygen will be discussed first.

1. Redox Systems Not Dependent on Surface Oxides
We saw that $Fe(CN)_6^{-3/-4}$, AA, and DA could exhibit fast ET
kinetics on GC surfaces with a surface O/C that was immeasurably
low as judged by XPS or surface voltammetry. Furthermore,
large variations in surface oxygen and observed capacitance
had little effect on observed kinetics. It is possible, in fact

likely, that a carbon surface exposed to electrolyte solution will
acquire some low level of surface oxygen, but the preponderance
of evidence indicates that k^o_{obs} for $Fe(CN)_6^{-3/-4}$, AA, and DA
can be high without any intentional oxidation. If surface oxygen
is involved in ET for these systems, it does so at much less
than monolayer coverage.

A secondary effect of surface oxides has been proposed to
explain the decrease in k^o for $Fe(CN)_6^{-3/-4}$ on polished GC as
the pH is increased from 4 to 7 [106]. At the higher pH, surface carboxylates should be deprotonated and will repel the
anionic $Fe(CN)_6^{-3/-4}$, leading to a local decrease in $Fe(CN)_6^{-3/-4}$
concentration near the surface. This Frumkin effect on k^o_{obs}
results in lower k^o_{obs} for $Fe(CN)_6^{-3/-4}$ and higher k^o for
$Ru(NH_3)_6^{+2/+3}$ at neutral pH. The effect is absent for laser-
activated surfaces, presumably because of the lower surface
O/C ratio [128]. Although this effect will alter k^o_{obs} for
$Fe(CN)_6^{-3/-4}$ and other ionic redox systems, the surface oxides
responsible for the effect are not required for high k^o_{obs}. Furthermore, the Frumkin effect is small (a factor of 2–5) compared
to the wide range of k^o values observed for various carbon
materials and preparation procedures.

Table 14 summarizes results for $Fe(CN)_6^{-3/-4}$ on various
electrode materials. Note that the highest rates for Au, Pt,
Pt/I, GC, and edge plane HOPG all fall in a range from 0.1–
0.5 cm/sec. Thus severe changes in surface composition, and
presumably double layer effects, seem to have little effect on
the $Fe(CN)_6^{-3/-4}$ rate. Although it is still possible that ET to
$Fe(CN)_6^{-3/-4}$ is inner sphere through a potassium ion, it seems
unlikely that K^+ would interact equally with widely variant surfaces. The results are consistent with an outer sphere ET to
$Fe(CN)_6^{-3/-4}$, with one or more K^+ ions associated with the
redox center.

TABLE 14

Observed k° for $Fe(CN)_6^{-3/-4}$ and Capacitance for Several Electrode Materials[a]

Electrode	Pretreatment	$k°_{obs}$ (cm/sec)	$C°_{obs}$ ($\mu F/cm^2$)	Ref.
Pt	$HClO_4$, KCl	0.24		100
Pt	flame	0.22		99
Pt/I	I adsorption	>0.1		103
Au	potential cycling	0.10		102
HOPG, basal	cleaved	$<10^{-7}$	2.0	31,92
HOPG, edge	laser	0.10	~70	92
HOPG, basal	laser, 50 MW/cm²	0.004	3.5	31,92,115
PG (0.5 M KCl)	cleaved	0.002		20
PG (0.5 M KCl)	ground	0.007		20
GC-20	VHT	0.14	10	59
GC-20	high speed polish	0.098		158
GC-20	ultraclean polish	0.14	~70	83
GC-20	conventional polish	0.005	24	126,159
GC-30	fractured	0.5	33	159
GC-30	polished, laser (25 MW/cm²)	>0.5	35	159
Graphite/hexane paste	none	0.003		146
Graphite/Nujol paste	none	1×10^{-4}		146

[a] 1 M KCl unless noted otherwise.

Whatever the specific role of K^+ in the $Fe(CN)_6^{-3/-4}$ reaction, it is clear that k^o is much faster on edge plane carbon.
Furthermore, AA and DA show similar anistropy of ET rate.
There are several possible reasons for the large basal/edge
anisotropy, including semiconductor character of basal plane,
higher wettability of edge plane, differential conductivity, etc.
Edge plane is more hydrophilic than basal plane, and ET to
aqueous redox systems may be much faster than on the hydrophobic basal plane. However, silanization of GC or PG had no
effect on ΔE_p for $Fe(CN)_6^{-3/-4}$ (see Table 8), even though the
electrode became hydrophobic. At this writing, there does not
appear to be an explanation for the edge/basal k^o anisotropy
which is consistent with all results.

The effect of the rate anisotropy on observed kinetics for
carbon electrode may be modeled, given a few assumptions.
This model should be considered a working hypothesis useful
for designing experiments and predicting carbon electrode behavior. Furthermore, it is based on experiments involving primarily $Fe(CN)_6^{-3/-4}$, AA, and DA, and its generality is not
yet clear. First, assume that any edge or basal domains on a
carbon surface are small compared to the diffusion layer thickness (case 1, Figs. 31 and 32). In this case, Eq. (24) applies
to k_{obs}^o. Second, assume that any surface roughness occurs on
a scale small compared to $\sqrt{Dt}$. In this case, the observed rate
is governed by the microscopic area, and

$$k_{obs}^o = \sigma \, k^o \tag{27}$$

where

$$\sigma = A_m/A_g$$

Combining with Eq. (24) yields Eq. (28):

$$k_{obs}^o = \sigma \, [k_{edge}^o \, f_e + k_{basal}^o \, (1 - f_e)] \tag{28}$$

Third, assume that $k^o_{basal} \ll k^o_{edge}$, as is true for DA and $Fe(CN)_6^{-3/-4}$. The second term in Eq. (28) is negligible except for very small f_e, and

$$k^o_{obs} = \sigma \, k^o_{edge} \, f_e \qquad (29)$$

Fourth, assume that some impurity adsorbs on the surface (edge and basal plane) with an average coverage Θ_p, and that adsorption on edge plane yields a smaller rate constant, k^o_p, then

$$k^o_{obs} = f_e \, \sigma \, [k^o_{edge} \, (1 - \Theta_p) + k^o_p \, \Theta_p] \qquad (30)$$

Finally, if $k^o_{edge} \gg k^o_p$:

$$k^o_{obs} = \sigma \, f_e \, k^o_{edge} \, (1 - \Theta_p) \qquad (31)$$

Θ_p might be comprised of trace organics from solution, polishing debris, or an oxidized carbon film which prevents ET. Although Eq. (31) is probably oversimplified, it does incorporate several variables which affect k^o_{obs} at carbon electrodes. For HOPG basal plane, f_e is near zero and k^o_{obs} is very slow ($<10^{-7}$ cm/sec). When HOPG basal plane is laser activated, the Raman spectrum indicates edge plane formation ($f_e > 0$) and a large rate increase is observed (10^{-7} to 0.004 cm/sec). GC is inherently rich in edge plane, and any GC surface will have a finite f_e. A crude estimate of $f_e = 10\%$ is available from capacitance and rate measurements [92]. However, preparation methods vary greatly in resulting surface cleanliness, leading to large variation of $1 - \Theta_p$. For fractured or VHT GC, Θ_p appears to be small and k^o_{obs} is large (0.14–0.5 cm/sec). Similarly, laser activation increases $1 - \Theta_p$ without significant changes in σ or f_e [159] and produces a GC electrode with high k^o and low C^o_{obs}. Since laser activation or VHT of GC cause large improvements in k^o_{obs} without accompanying changes in C^o_{obs} or f_e, changes in roughness or edge plane density do not appear to be responsible for activation. Stated more concisely, one must create edge plane sites on HOPG in order to increase k^o_{obs}, whereas polished GC has a high density of sites which must be cleaned.

Electrochemical activation of HOPG and GC is consistent with the model. ECP of HOPG causes lattice damage during oxidation, thus creating sites and increasing f_e. ECP on GC removes surface impurities and exposes or creates edge plane sites. Thus k^o enhancement for AA, DA, and $Fe(CN)_6^{-3/-4}$ does not require surface oxides, but formation of oxides indirectly increases k^o_{obs} by exposing or creating edge plane sites. Oxides can block the surface or cause electrostatic effects on k^o, but oxides do not appear to be required for fast ET kinetics for these systems.

2. Surface Oxide Mediated Redox Systems

There are clear examples where surface oxides do accelerate ET, such as cyt c, NADH oxidation, and O_2 reduction. In some cases, an H/D isotope effect has been observed, which was attributed to proton transfer to or from surface phenols. For these cases a clean surface is not sufficient for fast ET, and ECP is doing more than removing surface debris. Redox mediation (e.g., NADH), proton-assisted ET (e.g., EQ. (26)) and interaction with oxide binding sites (cyt c) are three examples where surface oxides are directly involved with ET. As noted by Kuwana [118], redox mediation may provide an alternate route for AA oxidation if "pristine" carbon sites are unavailable. Since each redox system in this class may have a distinct mechanism, it is difficult to generalize about oxide mediated ET, except to say that it will be slow in the absence of surface oxide, unlike $Fe(CN)_6^{-3/-4}$, AA, and DA.

B. Capacitance of Carbon Electrodes

As noted in Sec. IV, there are at least two, possibly three contributions to observed capacitance. With the exception of HOPG, there appears to be no correlation of C^o_{obs} with k^o_{obs}. In the absence of surface oxides, C_{obs}^o appears to be a weighted average of edge and basal contributions. Thus C_{obs}^o and k_{obs}^o on laser-damaged HOPG are correlated because they both depend on f_e [92]. Surface oxidation greatly increases C_{obs}^o due to surface faradaic reactions and perhaps the formation of micropores, but these processes do not necessarily affect k_{obs}^o. From an analytical standpoint, it is possible and usually desirable to make carbon surfaces with low C_{obs}^o and high k_{obs}^o.

C. Adsorption

Adsorption to carbon electrodes is best summarized by considering three different surfaces. First, an oxide-free surface is likely to exhibit differential adsorption on edge or basal regions, so the observed surface concentration will be a weighted average. Although it is well known that edge plane adsorbs more strongly than basal for gases, the effect has not been addressed electrochemically in detail to the author's knowledge. A second distinct type of carbon surface has monolayer or submonolayer coverage of surface oxides. Most polished surfaces, plasma-treated surfaces, etc., will fall into this group. Such surfaces have many oxide sites for adsorption, particularly of cations. Surface oxides may be responsible for the adsorption of DA observable at high scan rates, or the preferential ET to cations rather than anions on carbon films. The third surface is the heavily oxidized carbon electrode, which has a multilayer film of EGO. Multilayer adsorption of organic or inorganic cations occurs by ion exchange, thus preconcentrating cations from solution. This process results in major enhancement in sensitivity for cations, and suppression of anions. Although the film is not involved in ET per se, its adsorption properties have a major effect on electroanalytical response.

ACKNOWLEDGMENTS

The author acknowledges the hard work by the graduate students who contributed to the group's effort on carbon and who made the publications and progress in the area possible. In addition, many useful conversations with Mark Wightman, Ted Kuwana, Royce Engstrom, and their research groups are greatly appreciated. Finally, the laser activation and electrode kinetics work from our lab was supported by The Air Force Office of Scientific Research, and the development and application of Raman Spectroscopy to carbon materials was supported by the Surface and Analytical Chemistry Division of the National Science Foundation.

REFERENCES

1. R. C. Alkire, ed., New Horizons in Electrochemical Science and Technology, NMAB pub 438-1, National Academy Press, Washington, D. C., 1986.

2. K. Kinoshita, Carbon: Electrochemical and Physicochemical Properties, Wiley, New York, 1988.

3. S. Sarangapani, J. R. Akridge, and B. Schumm, eds., Proceedings of the Workshop on the Electrochemistry of Carbon, The Electrochemical Society, Pennington, NJ, 1984.

4. J.-P. Randin, in Encyclopedia of Electrochemistry of the Elements, Vol. 7, (A. J. Bard, ed.), Marcel Dekker, New York, 1976, pp. 1–291.

5. J. O. Besenhard and H. P. Fritz, Agnew Chem. Int. Ed. Engl. 22:950 (1983).

6. G. N. Kamau, Anal. Chim Acta. 207:1 (1988).

7. IUPAC Commission on Electroanalyt. Chem., M. Gross, and J. Jordan, Pure Appl. Chem. 56:1095 (1984).

8. G. Dryhurst, in Laboratory Techniques in Electroanalytical Chemistry, (P. T. Kissinger and W. R. Heineman, eds.), Marcel Dekker, New York, 1984.

9. J. Robertson, Adv. Phys. 35:317 (1986).

10. B. E. Warren, X-Ray Diffraction, Addison Wesley, New York, 1969.

11. F. Tuinstra, J. Koenig, J. Chem. Phys. 53:1126 (1970).

12. K. Kinoshita, Carbon: Electrochemical and Physiochemical Properties, Wiley, New York, 1988, p. 22.

13. M. S. Dresselhaus and G. Dresselhaus, Adv. Phys. 30:139 (1981).

14. P. A. Thrower, in Proceedings of the Workshop on the Electrochemistry of Carbon, The Electrochemical Society, Pennington, NJ, 1984, p. 40.

15. H. Ishida, H. Fukuda, G. Katagiri, and A. Ishitani, Appl. Spectros. 40:322 (1986).

16. N. Wada and S. A. Solin, Physica 105B:353 (1981).

17. M. W. Williams and E. T. Arakawa, J. Appl. Phys. 43: 3460 (1972).

18. R. J. Nemanich, G. Lucovsky, and S. A. Solin, Mater. Sci. Eng. 31:157 (1977).

19. R. J. Nemanich and S. A. Solin, Phys. Rev. B 20:392 (1979).

20. A. J. McQuillan and R. E. Hester, J. Raman Spec. 15:15 (1984).

21. M. Nakamizo, R. Kammereck, and P. L. Walker, Carbon 12:259 (1974).

22. M. Nakamizo, H. Honda, and M. Inagaki, Carbon, 16:281 (1978).

23. R. J. Nemanich, G. Lucovsky, and S. A. Solin, Solid State Comm. 23:117 (1977).

24. R. P. Vidano, D. B. Fischbach, L. J. Willis, and T. M. Loehr, Solid State Comm. 39:341 (1981).

25. R. Vidano and D. B. Fischback, J. Am. Cer. Soc. 61:13 (1978).

26. P. Lespade, A. Marchand, M. Couzi, and F. Cruege, Carbon 22:375 (1984).

27. T. Mernagh, R. Cooney, and R. Johnson, Carbon 22:39 (1984).

28. G. Katagiri, H. Ishida, and A. Ishitani, Carbon 26:565 (1988).

29. C. Underhill, S. Y. Leung, G. Dresselhaus, and M. S. Dresselhaus, Solid St. Comm. 29:769 (1979).

30. M. S. Dresselhaus and G. Dresselhaus, Adv. Phys. 30:290 (1981).

31. R. J. Bowling, R. T. Packard, and R. L. McCreery, J. Am. Chem. Soc. 111:1217 (1989).

32. A. Moore, in Chemistry and Physics of Carbon, Vol. 17 (P. L. Walker and P. A. Thrower, eds.), Marcel Dekker, New York, 1981, p. 233.

33. I. Spain, in Chemistry and Physics of Carbon, Vol. 16, (P. L. Walker and P. A. Thrower, eds.), Marcel Dekker, New York, 1981, p. 119.

34. M. S. Dresselhaus, G. Dresselhaus, K. Sugaihara, I. L. Spain, and H. A. Goldberg, Graphite Fibers and Filaments, Springer-Verlag, New York, 1988.

35. G. M. Jenkins and K. Kawamura, Polymeric Carbons, Carbon Fibre, Glass and Char. University Press, Cambridge, England, 1976.

36. A. Moore, in Chemistry and Physics of Carbon, Vol. 17 (P. L. Walker and P. A. Thrower, eds.), Marcel Dekker, New York, 1981, p. 42.

37. P. L. Walker, Carbon 10:369 (1972).

38. W. E. Chambers, Mech. Eng. 37 (1975).

39. G. Binnig, H. Rohrer, C. Gerker, and E. Werbel, Phys. Rev. Lett. 49:57 (1982).

40. G. Binnig and D. P. Smith, Rev. Sci. Inst. 57:1688 (1986).

41. I. Morcos and E. Yeager, Electrochim. Acta. 15:257 (1972).

42. J. P. Randin and E. Yeager, J. Electroanal. Chem. 36:257 (1972).

43. J. P. Randin and E. Yeager, J. Electroanal. Chem. 58:313 (1975).

44. A. Moore, in Chemistry and Physics of Carbon, Vol. 17 (P. L. Walker and P. A. Thrower, eds.), Marcel Dekker, New York, 1981, p. 15.

45. K. Kinoshita, Carbon: Electrochemical and Physicochemical Properties, Wiley, New York, 1988, Chapter 1.

46. Data from Refs. 8,32,34,47 and Tokai product literature.

47. S. Yamada, Defense Ceramic Information Center, report 68-2, DOD, 1968.

48. See, for example R. E. Panzer and P. R. Elving, Electrochim, Acta 20:635 (1975).

49. T. Nakajima, A. Mabuchi, and R. Hagiwara, Carbon 26: 357 (1988).

50. Y. Maeda, Y. Okemoto, and M. Inagaki, J. Electrochem. Soc. 132:2369 (1985).

51. J. O. Besenhard et al., Syn. Metals 7:185 (1983).

52. F. Beck and H. Krohn, Syn. Metals 7:193 (1983).

53. M. Inagaki, J. Mater. Res. 4:1560 (1989).

54. R. C. Engstrom and W. V. Strasser, Anal. Chem. 57:2759 (1985).

55. L. J. Kepley and A. J. Bard, Anal. Chem. 60:1459 (1988).

56. R. C. Engstrom, Anal. Chem. 54:2310 (1982).

57. A. P. Brown and F. C. Anson, Anal. Chem. 49:1589 (1977).

58. M. P. Soriaga and A. T. Hubbard, J. Am. Chem. Soc. 104:2736 (1982).

59. D. T. Fagan, I. F. Hu, and T. Kuwana, Anal. Chem. 57:2759 (1985).

60. G. Cabaniss, A. Diamantis, W. R. Murphy, P. W. Linton, T. J. Meyer, J. Am. Chem. Soc. 107:1845 (1985).

61. C. Kozlowski and P. M. D. Sherwood, J. Chem. Soc., Faraday Trans I 80:2099 (1984).

62. K. Kinoshita, Carbon: Electrochemical and Physicochemical Properties, Wiley, New York, 1088, p. 116.

63. J. Wandass, J. A. Gardella, N. Weinberg, M. Bolster, and L. Salvati, J. Electrochem. Soc. 134:2734 (1987).

64. C. Kozlowski and P. M. A. Sherwood, J. Chem. Soc., Farad Trans I 81:2745 (1985).

65. C. W. Miller, D. H. Karweik, and T. Kuwana, Anal. Chem. 53:2319 (1981).

66. M. Dautartas, J. F. Evans, and T. Kuwana, Anal. Chem. 51:104 (1979).

67. G. N. Kamau, W. S. Willis, and J. F. Rusling, Anal. Chem. 57:545 (1985).

68. T. P. Tougas and W. P. Collier, Anal. Chem. 59:2269 (1987).

69. W. G. Collier and T. P. Tougas, Anal. Chem.59:396 (1987).

70. A. Campion, J. Brown, and M. W. Grizzle, Surf. Sci. 115:L153 (1982).

71. H. Ishida et al., Appl. Spec. 40:322 (1986).

72. K. Kinoshita, Carbon: Electrochemical and Physicochemical Properties, Wiley, New York, 1988,

73. R. Ianniello, H. H. Wieck, and A. M. Yacynych, Anal. Chem. 55:2067 (1983).

74. J. Foley, C. Korzeniewski, J. Daschbach, and S. Pons, Electroanalytical Chemistry, Vol. 14 (A. J. Bard, ed.), Marcel Dekker, New York, 1988.

75. M. D. Porter, T. Bright, D. Allara, C. Chidsey, J. Am. Chem. Soc. 109:3559 (1987).

76. M. Porter, T. Bright, D. Allara, and T. Kuwana, Anal. Chem. 58:2461 (1986).

77. M. Schrader, J. Phys. Chem. 79:2508 (1975).

78. A. A. Gewirth and A. J. Bard, J. Phys. Chem. 92:5563 (1988).

79. J. Kwak and A. J. Bard, Anal. Chem. 61:1794 (1989).

80. R. Engstrom, M. Webber, D. J. Wunder, R. Burgess, S. Winquist, Anal. Chem. 58:844 (1986).

81. R. C. Engstrom and C. M. Pharr, Anal. Chem. 61:1099A (1989).

82. D. T. Fagan and T. Kuwana, Anal. Chem. 61:1017 (1989).

83. I. F. Hu, D. H. Karweik, and T. Kuwana, J. Electroanal. Chem. 188:59 (1985).

84. Y. Oren, H. Tobias, and A. Soffer, J. Electroanal. Chem. 162:87 (1984).

85. Y. Oren, H. Tobias, and A. Soffer, J. Electroanal. Chem. 186:63 (1985).

86. Y. Oren, H. Tobias, and A. Soffer, J. Electroanal. Chem. 206:101 (1986).

87. T. Nagaoka and T. Yoshino, Anal. Chem. 58:1037 (1986).

88. T. Nagaoka, T. Fukanaga, T. Yoshino, I. Wantanabe, T. Nakayama, and S. Okazaki, Anal. Chem. 60:2766 (1988).

89. T. Nagaoka, T. Sakai, K. Ogura, and T. Yoshino, Anal. Chem. 58:1953 (1986).

90. E. Gileadi and N. Tshernikovski, Electrochim. Acta 16: 579 (1971).

91. E. Gileadi, N. Tshernikovski, and M. Babai, J. Electroanal Chem. 119:1018 (1972).

92. R. Rice and R. L. McCreery, Anal. Chem. 61:1637 (1989).

93. D. T. Fagan, I. F. Hu, and T. Kuwana, Anal. Chem. 57: 2761 (1985).

94. D. Laser and M. Ariel, J. Electroanal. Chem. 52:291 (1974).

95. For example, see R. W. Murray, in Electroanalytical Chemistry, Vol. 13 (A. J. Bard, ed.), Marcel and Dekker, New York, p. 191 (1984).

96. A. J. Bard and L. R. Faulkner, Electrochemical Methods, Wiley, New York, 1980.

97. A. J. Bard, R. Parsons, and J. Jordan, Standard Potentials in Aqueous Solution, Marcel Dekker, New York, 1985, p. 408.

98. R. N. Adams, Electrochemistry at Solid Electrodes, Marcel Dekker, New York, 1969.

99. E. L. Goldstein and M. R. Van De Mark, Electrochim Acta 27:1079 (1982).

100. P. H. Daum and C. G. Enke, Anal. Chem. 41:653 (1969).

101. L. M. Peter, W. Durr, P. Bindra, and M. Gerisher, J. Electroanal. Chem. 71:31 (1976).

102. J. Kuta and E. Yeager, J. Electroanal. Chem. 59:110 (1975).

103. J. Kawaik, P. Kulesza, and Z. Galus, Anal. Chem. 226: 305 (1987).

104. R. Sohr and L. Muller, Electrochim. Acta 20:451 (1975).

105. J. Kawaik, T. Jedral, and Z. Galus, J. Electroanal. Chem. 145:163 (1983).

106. M. R. Deakin, K. J. Stutts, and R. M. Wightman, J. Electroanal. Chem. 182:113 (1985).

107. C. R. Martin, I. Rubenstein, and A. J. Bard, J. Electroanal. Chem. 151:267 (1983).

108. M. R. Deakin and R. M. Wightman, J. Electroanal. Chem. 206:167 (1986).

109. M. R. Deakin, P. Kovach, K. Stutts, and R. M. Wightman, Anal. Chem. 58:1474 (1986).

110. M. D. Hawley, S. Tatawawadi, S. Piekarski, and R. N. Adams, J. Am. Chem. Soc. 89:447 (1967).

111. A. W. Sternson, R. McCreery, B. Feinberg, and R. N. Adams, J. Electroanal. Chem. 46:313 (1973).

112. S. P. Perone and W. J. Kretlow, Anal. Chem. 38:1760 (1966).

113. Z. Vavrin, Coll. Czech. Chem. Comm. 14:367 (1949).

114. R. S. Nicholson and I. Shain, Anal. Chem. 36:706 (1964).

115. R. Wightman, M. Deakin, P. Kovach, W. Kuhr, and K. Stutts, J. Electrochem. Soc. 131:1578 (1984).

116. J. J. Ruiz, A. Aldaz, and M. Dominguez, Can. J. Chem. 55:2799 (1977).

117. J. J. Ruiz, A. Aldaz, and M. Dominguez, 56:1533 (1978).

118. I. F. Hu, and T. Kuwana, Anal. Chem. 58:3235 (1986).

119. C. Amatore, J. M. Saveant, and D. Tessier, J. Electroanal. Chem. 147:39 (1983).

120. F. A. Armstrong, A. M. Bond, H. A. O. Hill, I. S. M. Psalti, C. G. Zoski, J. Phys. Chem. 93:6485 (1989).

121. S. M. Lipkin, G. L. Cahen, G. E. Stoner, L. L. Scriber, and E. Gileadi, J. Electroanal. Chem. 135:368 (1988).

122. R. Bowling and R. L. McCreery, Anal. Chem. 60:605 (1988).

123. J. P. Randin and E. Yeager, J. Electrochem. Soc. 118:711 (1972).

124. K. Kinoshita, Carbon: Electrochemical and Physicochemical Properties, Wiley, New York, 1988, p. 294.

125. E. Hershenhart, R. McCreery, and R. Knight, Anal. Chem. 56:2256 (1984).

126. M. Poon and R. L. McCreery, Anal. Chem. 58:2745 (1986).

127. M. Poon and R. L. McCreery, Anal. Chem. 59:1615 (1987).

128. M. Poon, R. L. McCreery, and R. Engstrom, Anal. Chem. 60:1725 (1988).

129. R. Bowling, R. L. McCreery, C. M. Pharr, and R. C. Engstrom, Anal. Chem. 61:2763 (1989).

130. J. Steinbeck, G. Braunstein, M. Dresselhaus, T. Venkatesan, and D. Jacokson, J. Appl. Phys. 58:4374 (1985); 64:1802 (1988).

131. J. S. Speck, J. Steinbeck, G. Braunstein, M. S. Dresselhaus, and T. Venkatesan, Beam-Solid Interactions and Phase Transformations, Proceedings of the Materials Research Society, Pittsburgh, 1985.

132. J. S. Speck, J. Steinbeck, and M. S. Dresselhaus, J. Mater. Res. 5:980 (1990).

133. R. A. Wightman, E. C. Paik, S. Borman, and M. A. Dayton, Anal. Chem. 50:1410 (1978).

134. R. C. Engstrom and V. A. Strasser, Anal. Chem. 56:136 (1985).

135. R. Bowling, R. Packard, and R. L. McCreery, Langmuir 5:683 (1989).

136. R. E. Panzer and P. J. Elving, J. Electrochem. Soc. 119:864 (1972).

137. R. E. Panzer and P. J. Elving, Electrochim. Acta 20:635 (1975).

138. R. S. Niocholson, Anal. Chem. 37:1351 (1965).

139. F. A. Armstrong and K. J. Brown, J. Electroanal. Chem. 219:319 (1987).

140. C. M. Elliott and R. W. Murray, Anal. Chem. 48:1247 (1976).

141. F. A. Armstrong, A. M. Bond, H. A. O. Hill, I. S. Psalti, and C. G. Zoslei, J. Phys. Chem. 93:6485 (1989).

142. F. A. Armstrong, A. M. Bond, H. A. O. Hill, B. N. Oliver, and I. S. Psalti, J. Am. Chem. Soc. 111:9185 (1989).

143. R. N. Adams, Anal. Chem. 30:1576 (1958).

144. R. N. Adams, Electrochemistry at Solid Electrodes, Marcel Dekker, New York, 1969.

145. R. E. Shoup, High Performance Liquid Chromatography 4:91 (1986).

146. M. Rice, Z. Galus, and R. Adams, J. Electroanal. Chem. 176:169 (1984).

147. C. Urbaniczky and K. Lundstrom, J. Electroanal. Chem. 176:169 (1984).

148. D. E. Weisshaar, D. E. Tallman, and J. L. Anderson, Anal. Chem. 53:1809 (1981).

149. J. P. McLean, Anal. Chem. 54:1169 (1982).

150. J. L. Anderson, K. Whitten, J. Brewster, T-Y. Ou, and W. K. Nonidez, Anal. Chem. 57:1366 (1985).

151. J. E. Anderson, D. Hopkins, J. Shadrick, and Y. Ren, Anal. Chem. 61:2330 (1989).

152. B. R. Shaw and K. E. Creasy, Anal. Chem. 60:1241 (1988).

153. J. Park and B. R. Shaw, Anal. Chem. 61:B48 (1989).

154. L. Falat and H. Y. Cheng, J. Electroanal. Chem. 157:393 (1983).

155. L. Falat and H. Y. Cheng, Anal. Chem. 54:2108 (1982).

156. S. Pong and T. Kuwana, J. Electrochem. Soc. 131:813 (1984).

157. D. Thornton, K. Corky, V. Spendel, J. Jordan, A. Robbat, D. Rutstrom, M. Gross, and G. Ritzler, Anal. Chem. 57:150 (1985).

158. G. Kamau, W. Willis, and J. Rusling, Anal. Chem. 57:545 (1985).

159. R. Rice, N. Pontikos, and R. McCreery, J. Am. Chem. Soc. 112:4617 (1990).

160. J. Zak and T. Kuwana, J. Am. Chem. Soc. 104:5514 (1982).

161. K. Stutts, P. Kovach, W. Kuhr, and R. Wightman, Anal. Chem. 55:1632 (1983).

162. G. Hance and T. Kuwana, Anal. Chem. 59:131 (1987).

163. R. Rice, C. Allred, and R. McCreery, J. Electroanal. Chem. 263:163 (1989).

164. B. Kazee, D. Weisshaar, and T. Kuwana, Anal. Chem. 57:2739 (1985).

165. R. J. Taylor and A. A. Humffray, J. Electroanal. Chem. 42:347 (1973).

166. J. P. Randin and E. Yeager, J. Electroanal. Chem. 58:313 (1975).

167. L. Bodalbhai and A. Brajter-toth, Anal. Chem. 60:2557 (1988).

168. K. Stulik, J. Electroanal. Chem. 250:173 (1988).

169. R. Rice, M. S. thesis, Ohio State University, 1987.

170. K. Sternitzke, R. L. McCreery, C. Bruntlett, and P. T. Kissinger, Anal. Chem. 61:1989 (1989).

171. J. L. Ponchon, R. Cespuglio, F. Gonon, M. Jouvet, and J-F. Pujol, Anal. Chem. 51:1483 (1979).

172. F. Gonon, M. Buda, R. Cespuglio, M. Jouvet, and J. Pujol, Nature (London) 286:902 (1980).

173. A. Beilby and A. Carlsson, J. Electroanal. Chem. 248: 283 (1988).

174. K. M. Sundberg, W. H. Smyrl, L. Atanasoska, and R. Atanasoski, J. Electrochem. Soc. 136:434 (1989).

175. J. Wang and M. S. Lin, Anal. Chem. 60:499 (1988).

176. J. Wang and P. Tuzhi, Anal. Chem. 58:1787 (1986).

177. D. M. Anjo, Anal. Chem. 61:2603 (1989).

178. R. E. Vasquez, M. Hono, A. Kitani, and K. Sasaki, J. Electroanal. Chem. 196:397 (1985).

179. N. Cenas, J. Rozgaite, A. Pocius, and J. Kulys, J. Electroanal. Chem. 154:121 (1983).

180. D. Tse and T. Kuwana, Anal. Chem. 50:1315 (1978).

181. R. M. Wightman, L. May, and A. C. Michael, Anal. Chem. 60:769A (1988).

182. L. A. Knecht, E. J. Guthrie, and J. Jorgenson, Anal. Chem. 56:479 (1984).

183. J. G. White and J. W. Jorgenson, Anal. Chem. 58:2992 (1986).

184. J. Feng, M. Brazell, K. Renner, R. Kasser, and R. N. Adams, Anal. Chem. 59:1863 (1987).

185. J. Baur, E. Kristensen, L. May, D. Wiedemann, and R. M. Wightman, Anal. Chem. 60:1268 (1988).

186. R. Kelly and R. M. Wightman, Anal. Chem. Acta 187:79 (1986).

187. Ronald Rice and R. L. McCreery, unpublished.

188. A. C. Michael and J. B. Justice, Anal. Chem. 59:1863 (1987).

189. A. Beilby and A. Carlsson, J. Electroanal. Chem. 248: 283 (1988).

190. S. Sujaritvanichpong, K. Aoki, K. Tokuda, and H. Matsuda, J. Electroanal. Chem. 198:195 (1986).

191. F. G. Gonon, C. M. Fombarlet, M. J. Buda, and J. F. Pujol, Anal. Chem. 53:1386 (1981).

192. P. Kovach, M. Deakin, and R. M. Wightman, J. Phys. Chem. 90:4612 (1986).

193. R. A. Saraceno and A. G. Ewing, Anal. Chem. 60:2016 (1988).

194. Y-T. Kim, P. Sarnulis, and A. G. Ewing, Anal. Chem. 58:1782 (1986).

195. R. A. Saraceno, C. Engstrom, M. Rose, and A. G. Ewing, Anal. Chem. 61:560 (1989).

196. A. Rojo, A. Rosenstratten, and D. Anjo, Anal. Chem. 58:2988 (1986).

197. B. Hepler, S. Weber, and W. Purdy, Anal. Chem. Acta 102:41 (1978).

198. B. Hepler, S. Weber, and W. Purdy, Anal. Chem. Acta 113:269 (1980).

199. K. Lundstrom, Anal. Chem. Acta 146:97 (1983).

200. W. J. Blaedel and G. A. Mabbott, Anal. Chem. 50:933 (1978).

Author Index

Numbers in parentheses are reference numbers and indicate that an author's work is referred to although his name is not cited in the text. Underlined numbers give the page on which the complete reference is listed.

375

Subject Index

Adsorbate processes
 carbon monoxide as an adsorbate,
  on-line MS study of, 199–
  201
 ethanol adsorbates on platinum,
  on-line MS study of, 202–
  205
Adsorption/desorption of surfactant molecules, application
  of EQCM to, 47–53
AT-cut QCM crystals, 4–19
 two predominant types of, 8–10
Auger electron spectroscopy (AES)
  for examining carbon electron surface, 269–273

Carbon black, 250
Carbon composite electrodes, 322–
  323
Carbon dioxide, electroreduction
  of,
 on-line MS study of, 208–211
Carbon electrodes, structural
  effects of (see Structural
  effects of carbon electrodes)
Carbon fiber electrodes, preparation and performance of,
  345–353
 electrochemical pretreatment,
  348–353
Carbon fibers, 246–250
Carbon film electrodes, 353–357
Carbon films, 250–255

Carbon monoxide as an adsorbate,
  on-line MS study of, 199–201
Carbon paste (CP) electrodes, 320–
  322
Cells, method of mounting QCM
  crystals to, 13–14
Chemisorbed oxygen (carbon electrode surface structural variable), 258–266
Composite polymer films, EQCM
  applications to, 74–81
Conducting polymers, EQCM applications to, 70–81
Copolymers, EQCM applications to,
  74–81

Differential electrochemical mass
  spectroscopy (DEMS), 183
Diheptylviologen bromide (DHVBr),
  electrochromic films of, deposition and dissolution of, 54–
  58
Distribution of edge vs. basal plane
  on surface (carbon electrode
  surface structural variable),
  256–257

Electrochemical characterization of
  carbon electrodes, 280–302
 adsorption, 301–302
 electron transfer kinetics, 287–
  301

389